ENERGY METABOLISM IN INSECTS

ENERGY METABOLISM IN INSECTS

Edited by

Roger G. H. Downer

University of Waterloo
Waterloo, Ontario, Canada

Springer Science+Business Media, LLC

Library of Congress Cataloging in Publication Data

Main entry under title:

Energy metabolism in insects.

Includes papers presented at a symposium held during the 16th International Congress of Entomology in Kyoto, Japan, Aug. 3-9, 1980.

Includes bibliographies and index.

1. Insects—Physiology—Congresses. 2. Energy metabolism—Congresses. I. Downer, Roger G. H., 1942- . II. International Congress of Entomology (16th : Kyoto, Japan)

QL495.E49 595.7'0133 81-11839

ISBN 978-1-4615-9223-5 ISBN 978-1-4615-9221-1 (eBook) AACR2

DOI 10.1007/978-1-4615-9221-1

Originally published by Plenum Press, New York in 1981

Softcover reprint of the hardcover 1st edition 1981

A Division of Plenum Publishing Corporation

233 Spring Street, New York, N.Y. 10013

Contributors

A.M.T. Beenakkers • Laboratory of Chemical Animal Physiology, State University of Utrecht, 3508 TB Utrecht, The Netherlands

Einar Bursell • Department of Zoology, University of Rhodesia, Salisbury, Rhodesia. Present address: Tsetse Research Laboratory, Department of Veterinary Medicine, Langford, near Bristol, England

David J. Candy • Department of Biochemistry, University of Birmingham, Birmingham, United Kingdom

Haruo Chino • Biochemical Laboratory, Institute of Low Temperature Science, Hokkaido University, Sapporo, Japan

R.G.H. Downer • Department of Biology, University of Waterloo, Waterloo, Ontario, Canada

L.L. Keeley • Department of Entomology, Texas Agricultural Experiment Station, Texas A&M University, College Station, Texas

J.E. Steele • Department of Zoology, University of Western Ontario, London, Ontario, Canada

R.H.C. Strang • Department of Biochemistry, University of Glasgow, Glasgow, Scotland

D.J. Van der Horst • Laboratory of Chemical Animal Physiology, State University of Utrecht, 3508 TB Utrecht, The Netherlands

W.J.A. Van Marrewijk • Laboratory of Chemical Animal Physiology, State University of Utrecht, 3508 TB Utrecht, The Netherlands

Preface

The scientific program for the XVI International Congress of Entomology, held in Kyoto, Japan August 3–9, 1980 included a symposium on the subject of "Energy Metabolism and Its Regulation in Insects." The symposium provided an opportunity to integrate knowledge, and focus attention, on an important and fundamental aspect of insect biochemistry/physiology. The energy metabolism of insects differs from that of other animals in a variety of ways, including the prodigious amounts of energy expended by flying insects, the presence in hemolymph of large concentrations of sugar in the form of the nonreducing disaccharide trehalose, the transport of fat in the form of diacylglycerol, and the periodic mobilization and deposition of cuticular components during development. These differences, together with hormones, neurohormones, and neurotransmitters that are specific to (or functionally different in) insects, serve to demonstrate the unique nature of energy metabolism in insects. An obvious corollary from the demonstrated uniqueness of insect energy metabolism is that an understanding of the process may lead to the development of new, specific agents or strategies for the suppression of insect pests.

The present volume is an expanded version of the Kyoto symposium. It includes contributions from authors who were unable to attend the original symposium and provides an account of our current understanding of energy metabolism in insects. The first chapter introduces the subject and indicates environmental and physiological factors that effect changes in the basic pattern of energy flow. Chapter 2 identifies the chemical factors responsible for mediating changes in energy metabolism and describes the action of these factors in regulating the mobilization, transport, and utilization of energy-rich substrates. The next three chapters deal with the roles of carbohydrate, lipid, and proline respectively in fulfilling the energy requirements of particular tissues under particular physiological conditions. Chapter 6 describes the nature and role of the specific lipoprotein responsible for transport of diacylglycerols and other lipids in insects. Chapter 7 considers supply and utilization of energy in the insect nervous system and compares these processes with energy supply in other tissues and organisms. The final chapter describes the properties, nature, and functional development of the mitochondrion.

In a volume of this nature it is inevitable that some redundancy occurs between certain contributions; such overlap is not necessarily a negative feature and indeed may serve to enhance the volume by permitting a more balanced and integrated treatment of the subject. A more serious concern arises through omissions and/or inappropriate depth of treatment given to particular topics; these constraints are occasioned in part by the withdrawal of some invited participants, the individual interests and biases of contributors, and, undoubtedly, oversight on the part of the editor. It is hoped that any such shortcomings will not detract seriously from the usefulness of the volume, and that the book will stimulate interest in a relatively neglected area of insect biochemistry/physiology.

It is a pleasure to acknowledge the assistance that I have received from several individuals in bringing this volume to press. Professor Haruo Chino and the organizing committee of the XVI International Congress of Entomology proposed the symposium and provided the impetus that was required to initiate the project. Kirk Jensen of Plenum Press has provided valuable counsel in the organization and production of the volume. Finally, I express my appreciation to the individual authors for their enthusiastic cooperation in producing excellent manuscripts that included many unpublished observations. Any success that the book enjoys can be attributed to their efforts.

R.G.H. Downer

Waterloo, August 1980

Contents

Chapter 3

Role of Lipids in Energy Metabolism

A.M.T. Beenakkers, D.J. Van der Horst, and W.J.A. Van Marrewijk

Chapter 4

The Role of Carbohydrate Metabolism in Physiological Function

J.E. Steele

Chapter 5

The Role of Proline in Energy Metabolism

Einar Bursell

Chapter 6

Lipid Transport by Hemolymph Lipoprotein—A Possible Multiple Role of Diacylglycerol-Carrying Lipoprotein

Haruo Chino

Chapter 7

Energy Metabolism in the Insect Nervous System

R.H.C. Strang

Chapter 8

Neuroendocrine Regulation of Mitochondrial Development and Function in the Insect Fat Body

L.L. Keeley

1

Physiological and Environmental Considerations in Insect Bioenergetics

R.G.H. Downer

1. Introduction

Energy and the ability to use it in a controlled, programmed manner constitute the most basic requirement and property of life. Therefore, bioenergetics—the study of the energy transformations that occur within living organisms (Lehninger, 1971)—is an important and fundamental aspect of the biology of any living system. The present volume is concerned primarily with molecular/biochemical and physiological aspects of bioenergetics in insects, a class of animals in which a multifarious array of bioenergetic processes are represented. Contributors to the volume will discuss the mobilization, transport, and utilization of energy-rich substrates and the various hormones, neurohormones, and neurotransmitters that serve to regulate these biochemical events. However, it is important to recognize that these diverse reactions do not occur in isolation; rather, they are essential components in a complex of integrated biochemical pathways that respond to particular environmental and physiological conditions so as to promote the biological success of the individual and the species. Thus, although it is inevitable that specific biochemical reactions are studied as independent entities, it is important that such reactions are also considered within a broader frame of reference.

R.G.H. Downer • Department of Biology, University of Waterloo, Waterloo, Ontario, Canada.

It is the objective of the present chapter to discuss in a holistic manner the concept of energy flow within the insect, and to identify certain environmental and physiological factors that influence the nature and direction of energy flow. From such considerations it is hoped that a perspective will be provided within which the remaining contributions may be viewed.

2. Energy Flow

The basic equation for the exchange of energy between an insect and its environment is:

$$E_{in} = E_{out}$$

where E_{in} represents the total energy acquired by an insect during its lifetime, and E_{out} describes the total energy expenditure of that insect (Carlson and Hsieh, 1970). A variety of factors contribute to this equation, and these are indicated in the generalized bioenergetic scheme illustrated in Fig. 1 and discussed in general terms below.

2.1. Sources of Acquired Energy

Insects obtain most of their energy in the form of chemical bonds within the carbohydrate, fat, and protein molecules that comprise the food consumed by the insect and the reserves that are contributed by the mother. Energy is released from these molecules by a progressive series of chemical reactions in which chemical bonds are cleaved and electrons undergo transitions from higher to lower energy levels. Liberated electrons are captured by electron acceptors and passed along a chain of electron carriers before finally reducing oxygen to water. Each successive reductive and oxidative reaction in the electron transport chain results in the electron's occupying a lower energy level, and the sequence is coupled to the formation of particular chemical bonds in adenosine triphosphate (ATP). Hydrolysis of the terminal phosphate of 1 mole of ATP provides a free energy yield of 7.3 kcal, and it is energy from this source that is used ultimately to perform biological work.

In addition to chemical energy, many insects absorb solar energy, which is used to maintain body temperature at optimal levels for metabolic function (Section 4.2), and the energy of light and sound is converted to electrical energy within specialized nervous receptors.

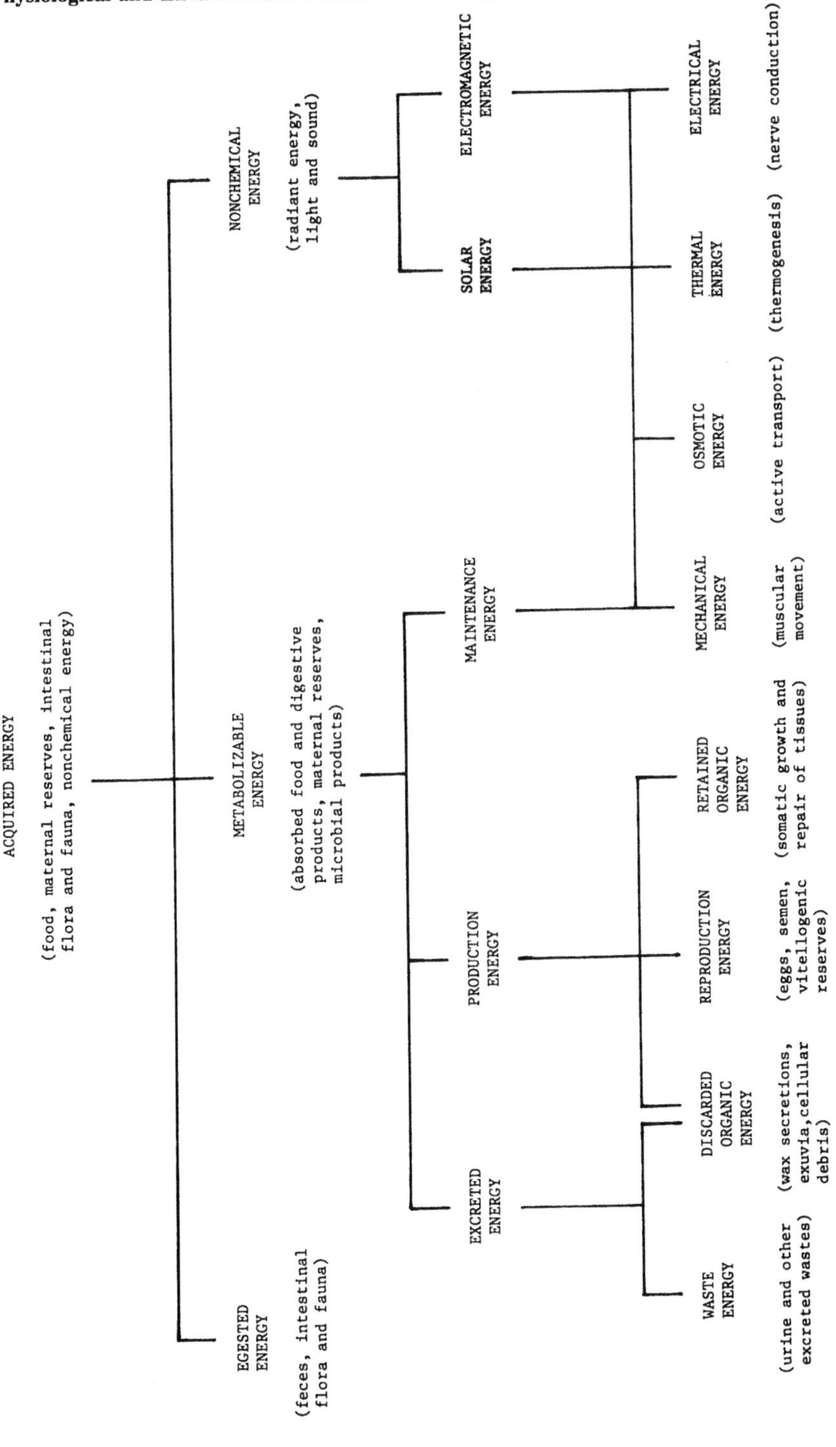

Fig. 1. Components in energy budget of insects.

2.2. Fate of Acquired Energy

Part of the total energy acquired by an insect is not absorbed across the intestinal tract, but is eliminated as nonabsorbed food or as assimilated products of intestinal flora and fauna. The total energy that is made available to the cells to perform biological work comprises the chemical energy absorbed across the intestinal tract and present in maternal reserves together with nonchemical energy acquired from the physical environment. This total energy is used primarily for cellular maintenance and to synthesize the diversity of organic molecules that are essential for growth and normal physiological function. The use of metabolizable energy for production and maintenance involves the original chemical energy undergoing transformation to form new chemical bonds, electrical energy, mechanical energy, osmotic energy, and thermal energy. In accordance with the second law of thermodynamics, each transformation results in loss of free energy and increased entropy of the environment. Therefore, there is continual transfer of energy throughout the life of the insect, and this energy must be replaced so that the insect can maintain a positive energy balance (Section 4.1). Figure 1 demonstrates also that some energy is lost from the insect in the form of waste products of metabolism and discarded biosynthetic products.

3. Interspecific Considerations in Bioenergetics

Insects, like most multicelled organisms, are able to store reserves of metabolizable energy. The accumulation of such reserves enables the insect to survive periods of famine and to engage in nonfeeding activities. The interrelationships that exist between reserves and the various components of metabolisable energy are illustrated in Fig. 2. Each arrow on the diagram may be considered as a vector that is subject to quantitative

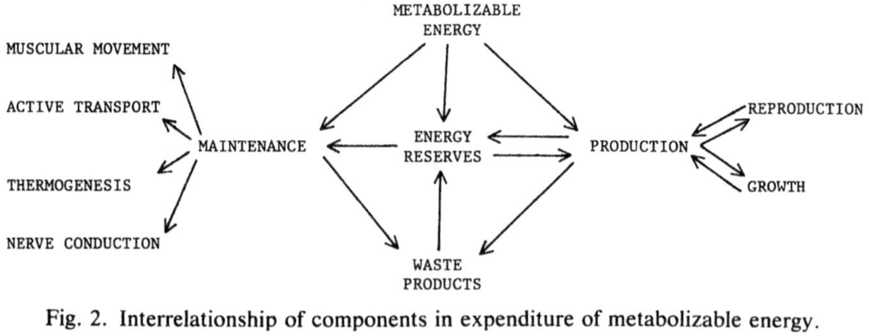

Fig. 2. Interrelationship of components in expenditure of metabolizable energy.

change according to the species under investigation and the particular physiological and ecological conditions that prevail at that time. The principal uses for metabolizable energy are production and maintenance, and factors that contribute to interspecific differences in the relative distribution of energy between these components are discussed below.

3.1. Production

The success of any species depends upon the ability of each generation to produce the optimal number of offspring, and the number of offspring produced reflects the reproductive strategy of the species. The reproductive strategy of an organism may be described as "r strategy" or "K strategy," referring to whether the organism produces the largest possible number of progeny (r) or offspring with the highest possible likelihood of survival (K) (MacArthur and Wilson, 1967). Thus, insects that direct all available energy into reproduction and then die represent an extreme example of "r selection," whereas insects that lay a single egg in an ample and protected food supply occupy the opposite end of the r–K continuum. It is apparent that the reproductive strategy will influence energy flow, with r strategists directing a major proportion of available reserves towards reproduction, whereas K strategists use a greater percentage of reserves for maintenance, intraspecific competition, and somatic growth. It should be noted that specific strategies are subject to variation or modification according to the conditions that prevail during a particular period. For example, reabsorption of oocytes and consequent redistribution of reproductive reserves among somatic tissues may occur in response to such factors as nutritional deficiency, virginity (or absence of males), chemical influences (such as queen substance in honeybees), and parasitism (Bell and Bohm, 1975).

3.2. Maintenance

Insects are aerobic organisms, and, therefore, the percentage of maintenance energy used to maintain the insect in a resting, undisturbed condition may be equated with basal oxygen consumption. Extrapolating from numerous studies on oxygen consumption in a variety of species (Schmidt-Nielsen, 1980), the relationship between basal maintenance energy and body size may be expressed by the exponential formula:

$$E_m = kW^b \text{ (or } \log E_m = b \log W + \log k)$$

where E_m is the total basal maintenance energy, W is the body weight, and k and b are constants. However, as b is usually less than 1, and often

0.75, the metabolic rate (oxygen consumption or energy expenditure per unit body weight) is greater in small insects than in large ones. Thus, the size of an insect determines the proportion of acquired energy that is available for purposes other than basal cellular maintenance. The size of an insect is a species-specific function determined primarily by the genetic constitution of the species, although the nutritional and developmental history of the individual also contribute to the final size. It should be emphasized that the above discussion is concerned only with energy required for basal cellular maintenance. Obviously, the behavioral patterns of different species will also influence the apportionment of energy reserves. For example, highly active insects will demonstrate a greater flow of metabolizable energy to the muscular movement component of maintenance than will be observed with relatively sedentary species. Finally, it should be noted that interspecific differences exist with regard to enzyme complements and the relative importance of particular energy-rich substrates in satisfying bioenergetic demands.

4. Factors Influencing Bioenergetic Flow in Individuals

The foregoing discussion pertains to the energy exchanges that occur during the entire life of an insect and that represent the sum of the many bioenergetic reactions that are proceeding at any particular moment. It is apparent that the metabolic state of an insect may range between extremes of biosynthesis/accumulation of reserves and the utilization of these reserves. Such variation results in marked differences in the magnitude and direction of metabolic flow between the various components identified in Fig. 2 during different physiological states. The bioenergetic condition of an insect is dynamic and influenced by a variety of factors, some of which are identified and discussed below.

4.1. Development

Insect growth occurs in a cyclical fashion, with periods of active growth alternating with relatively passive or nonactive stages. Superimposed on this cyclical pattern are a series of molts, during which there may be considerable cellular breakdown and resynthesis. These considerations, together with the allometric nature of growth in insects, suggest that major differences may occur in the pattern of bioenergetic flow between different stages of development (see also the chapter by Beenakkers in this volume.) The change in energy flow during the development of a holometabolous insect, *Tribolium castaneum,* is illustrated in Fig. 3 (Klekowski and Duncan, 1975; Prus, 1975). The diagram demonstrates that at various times during the life of *Tribolium*, the insect may be in a state of negative energy balance (Fig. 3a,b; $E_{in} < E_{out}$) or of positive energy balance (Fig. 3d; $E_{in} > E_{out}$) or may show a balanced energy budget (Fig. 3c; $E_{in} = E_{out}$). Under conditions of negative energy balance, accumulated reserves will be used to provide the energy for maintenance and production, whereas a state of positive energy balance will result in the accumulation of reserves. Quantitative differences in energy flow occur within the broad developmental categories identified in Fig. 3; for example, the curve of metabolic rate against pupal age is characteristically U-shaped for many holometabolous insects (Prosser, 1973). Developmental changes in energy flow are less pronounced in hemimetabolous insects, although some fluctuations may be anticipated around the time of the molt as cuticular resorption and deposition proceed (Lipke *et al.*, 1965). Developmental differences in insect metabolism have been described for many species (Chen, 1971; Agrell and Lundquist, 1973), and the elegant studies of L.L. Keeley and his co-workers (see chapter by Keeley in this volume) have demonstrated age-related changes in mitochondrial function. These studies serve to emphasize the importance of defining precisely the developmental stage and age of insects used in any study of energy metabolism.

It is appropriate also to indicate that when insects reach the adult stage the sex of the insect becomes an important factor in bioenergetic processes. An obvious factor that distinguishes the bioenergetic flow of females from that of males is the process of vitellogenesis, by which females transfer metabolic reserves to developing eggs to serve as substrates for growth and provision of energy during embryogenesis (see

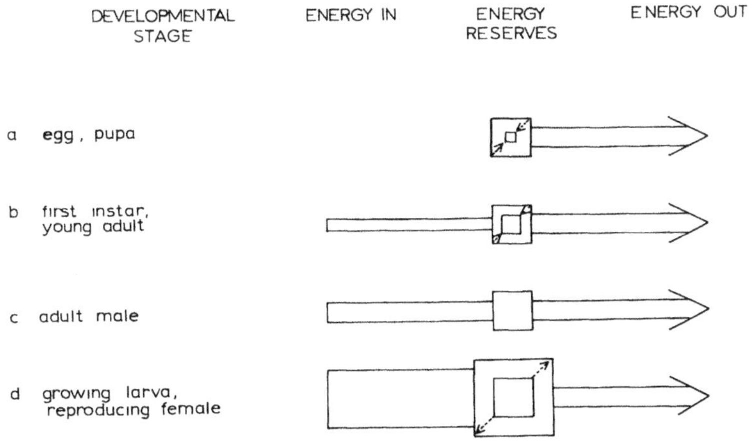

Fig. 3. Energy flow at specific stages in development of *Tribolium castaneum* (according to Klekowski and Duncan, 1975).

chapter by Beenakkers in this volume.) Pronounced sexual dimorphism in lipid content has been reported for adults of many species (Gilbert, 1967; Fast 1970), and these reports lend further support to the tenet that bioenergetic flow differs between male and female adult insects.

4.2. Temperature

Although insects are often considered to be true poikilotherms, there is considerable evidence to suggest that most species are capable of some degree of thermoregulation (May, 1979). Insects achieve thermoregulation either by generating metabolic heat or by varying the extent of heat exchange with the environment (e.g., by microhabitat selection). Metabolic heat arises as a result of part of the chemical energy released in metabolic reactions being dissipated as heat; it is this energy that maintains the body temperature of insects above that of the environment. The high metabolic rate of actively metabolizing flight muscles results in the generation of much heat and consequent elevation of thoracic temperature. The increase in thoracic temperature that occurs following the initiation of flight in the American cockroach is illustrated in Fig. 4, and these data demonstrate also a fairly rapid recovery and return to normal resting temperatures when flight ceases. Similar observations have been made for other insect species, and these studies indicate that the rise in temperature is restricted to the thoracic region (Heinrich and Bartholomew, 1971; Heinrich, 1975; May, 1979). The relatively high temperatures attained within the thorax of flying insects are required so that metabolic reactions may proceed at the high rates needed to satisfy the prodigious energy requirements of active flight muscles (Sacktor, 1970). The rate of a reaction increases with temperature according to the Q_{10} approximation:

$$Q_{10} = \left(\frac{K_1}{K_2}\right)^{10/t_1 - t_2}$$

where Q_{10} is the factor by which the rate of a reaction is increased for a rise of 10°C, and K_1 and K_2 are the reaction velocity constants at temperatures t_1 and t_2 respectively. The close relationship between temperature and the rate at which metabolic energy may be generated results in many species raising their thoracic temperature before taking off for flight. Preflight elevation of thoracic temperature is achieved either by behavioral strategies that facilitate the absorption of radiant energy from sunlight or by the generation of metabolic heat. Metabolic heat may be produced by "warm-up" muscular contractions (equivalent to shivering thermogenesis in vertebrates) or by specific heat-generating metabolic activities such as substrate cycling. Figure 5 illustrates the conversion of

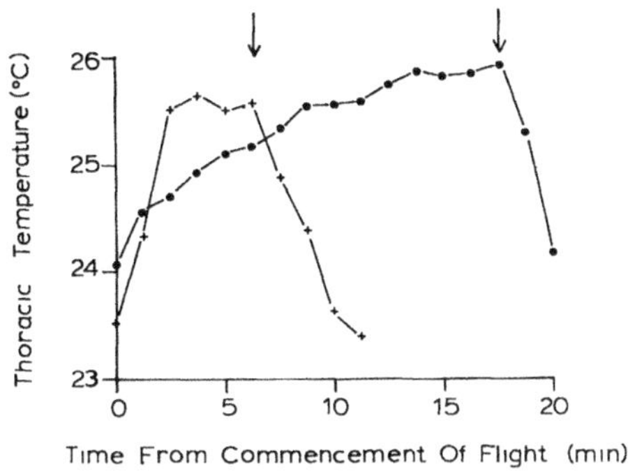

Fig. 4. Thoracic temperature of adult male cockroach during and immediately following flight. Ambient temperature = 21.85°C. Arrows indicate cessation of flight. (Temperature measured by an iron–constantan thermocouple inserted into thorax.)

fructose-6-phosphate to fructose-1,6-diphosphate in the glycolytic pathway, and it is evident that, in this conversion, catalyzed by phosphofructokinase, heat is generated at the expense of ATP. The reverse reaction is catalyzed by the enzyme fructose-1,6-diphosphatase. Both enzymes are present in flight muscle of bumblebees, but the extent of substrate cycling is greater at low temperatures than at high temperatures (Newsholme *et al.*, 1972). Therefore, at low temperatures the substrate cycle is operating, additional ATP is hydrolyzed, and heat is generated. It has been suggested that the operation of this cycle explains the ability of bumblebees to collect nectar at low ambient temperatures, whereas related species that do not employ the substrate cycle are relatively inactive at low temperatures.

Many studies have attempted to demonstrate a relationship between environmental temperature and the degree of saturation of fatty acids (Fast, 1970). The melting point of fatty acids increases with the degree of saturation, and, on this basis, it is suggested that there is likely to be a greater proportion of unsaturated to saturated fatty acids in insects raised at low temperatures. Unfortunately, most investigations have concentrated on the total fatty acid complement of insect tissues rather than on the membrane lipids, in which the maintenance of constant fluidity is of considerable importance. The membrane lipids represent a small proportion of the total lipid within most cells, and certainly within insects; therefore, changes that are restricted to this component are unlikely to

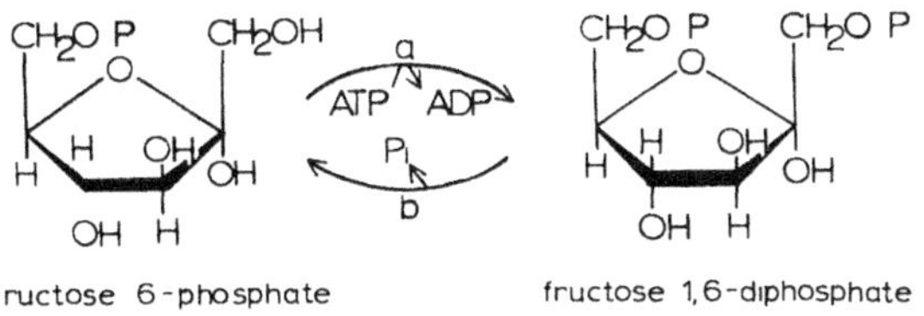

Fig. 5. Fructose-6-phosphate/fructose-1,6-diphosphate substrate cycle. a: Phosphofructokinase. b: Fructose-1,6-diphosphatase.

be revealed by examination of total lipids. Recent studies have examined the nature of mitochrondrial membranes taken from the flight muscles of locusts reared at 30°C and 45°C (V.L. Kallapur and R.G.H. Downer, unpublished observations). The results of these investigations suggest a temperature-induced change in membrane lipids, and studies are currently in progress to elucidate the nature and form of this change. Differences in the nature and/or composition of the membrane are likely to influence membrane fluidity and permeability; therefore, it is probable that environmental temperature, in addition to influencing metabolism according to the Q_{10} relationship, may affect the rate of substrate transport across biological membranes.

4.3. Excitation

An important, but frequently neglected factor in considering the physiological condition of experimental insects is the degree of excitation. Adult cockroaches demonstrate a characteristic physiological/biochemical response to handling that includes rapid stimulation of glycogenolysis in fat body, with concomitant elevation of hemolymph trehalose levels. The excitation-induced hypertrehalosemic response (EXIT-response) occurs in normal and head-ligated insects and is probably mediated by a neuroexcitatory factor released from the nervous system (Downer, 1979). Release of this factor converts the resting insect from a state of carbohydrate flux that favors the synthesis and deposition of glycogen to a trehalogenic condition in which glycogenolysis and trehalogenesis predominate (Downer, 1981). The significance of the metabolic transition is demonstrated by tracing the fate of [U-^{14}C]glucose under "excited" and "resting" conditions. When a small amount of radiolabeled glucose is injected into the cockroach hemocoel, the label appears rapidly as hemolymph trehalose owing to the EXIT-response induced by handling and injection; however, in resting insects, most excess hemolymph glucose is converted to glycogen (Spring *et al.*, 1977). The neurosecretory factor responsible for mediating the EXIT-response has not been positively identified, but there is evidence to suggest that it may be the monohydroxy-phenolamine compound, octopamine (Downer, 1980). Octopamine activates adenylate cyclase in a number of insect systems (Nathanson and Greengard, 1973; Robertson and Steele, 1973; Nathanson, 1979; Gole and Downer, 1980), and release of the amine may be expected to cause pronounced, short-term changes in metabolic flux. Octopamine occurs in all insects that have been examined for its presence (Robertson and Juorio, 1977), and, if this compound is the mediator of the excitation response in a manner that is analogous to the action of epinephrine in vertebrates, it is likely that excitation-induced effects on metabolism occur widely in insects.

Often, an associated consequence of excitation is increased muscular activity. Aspects of the utilization of energy reserves by thoracic musculature will be discussed by other contributors to this volume (see chapters by Candy, Beenakkers, Steele, and Strang) and will not be elaborated upon further at this stage. However, it is appropriate to identify an activity-related change in hemolymph trehalase that has been described in *P. americana*. As illustrated in Fig. 6, hemolymph trehalase is highly pH-sensitive, with a marked increase in enzyme activity evident between pH 7.5 and 6.5 (Matthews *et al.*, 1976); comparison of hemolymph pH from resting and recently flown insects demonstrates a pH range of 7.3–7.8 and 6.5–6.8, respectively, for each condition (Van Asperen and Van Esch, 1956; Downer and Matthews, 1977). Thus, the increased acidity of hemolymph that occurs during flight in the American cockroach results in increased trehalase activity; it is suggested that the increased hydrolysis of trehalose that occurs under these conditions facilitates the uptake of carbohydrate substrate by active flight musculature. The regulation of trehalase activity has not yet been fully elucidated (Katagiri, 1977; Van

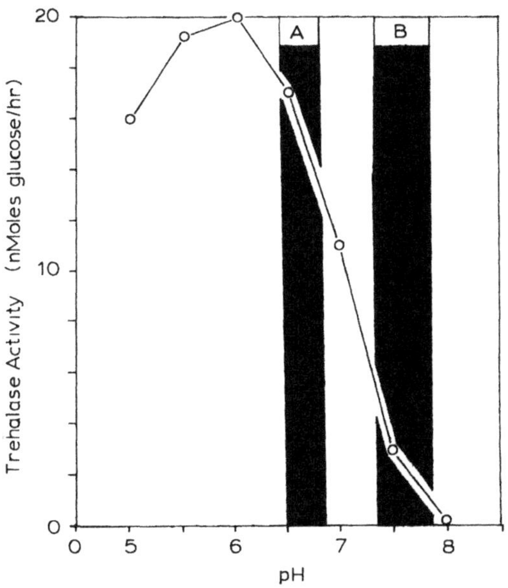

Fig. 6. Effect of pH on trehalase activity of cockroach hemolymph A: pH range of hemolymph from recently flown insects. B: pH range of hemolymph from rested insects (Matthews *et al.*, 1976; Downer and Matthews, 1977).

Handel, 1978; Downer and Matthews, 1978); however, it is evident that the several factors that appear to influence the activity of the enzyme affect carbohydrate flux in the insect.

4.4. Nutritional State

The ability of insects to feed periodically and to rely on accumulated reserves for their energy requirements during nonfeeding periods was indicated in Section 3. As a result of the cyclical feeding habit the metabolic state of many insects alternates between an absorptive phase, during which digestive products enter the hemolymph from the gut, and a postabsorptive phase, in which accumulated reserves are mobilized to provide the energy required by the insect. The frequency of feeding and the time taken for absorption of digestive products varies among species and according to the developmental stage. Such innate variation, together with a paucity of experimental data, render generalizations about this aspect of metabolic regulation in insects to be tenuous. In order to provide a basis for further discussion, it is useful to examine the differences between absorptive and postabsorptive metabolism in vertebrates, and these are summarized in Fig. 7. In the absorptive state, vertebrates utilize glucose as the principal energy source, with most newly absorbed amino acids and fats undergoing transformation into storage reserves; by contrast, in the postabsorptive state, gluconeogenic pathways generate glucose for utilization by the nervous system, but most of the animal's energy requirements are met by the oxidation of fat reserves. An indication of the postabsorptive metabolic state in insects may be obtained by considering the metabolic consequences of prolonged starvation. In the migratory locust, *Locusta migratoria,* starvation results in increased production of ketone bodies but decreased utilization of these compounds by fat body and flight muscles (Hill *et al.,* 1972). Ketone bodies provide an important source of energy for the vertebrate nervous system during prolonged starvation, but, unfortunately, no data are available to indicate if ketone bodies serve a similar glucose-sparing role in starved insects (see chapter by Strang in this volume.) Elevated hemolymph amino acid levels are also associated with starvation in the locust (Hill *et al.,* 1966), indicating mobilization of protein and the possibility of increased gluconeogenesis. Starved locusts maintain and elevate hemolymph diacylglycerol levels (Jutsum *et al.,* 1975; Mwangi and Goldsworthy, 1977), and a hyperlipidemic effect has also been described in larvae of the wax moth, *Galleria mellonella* (Wlodawer and Wiesniewska, 1965). Vertebrate plasma fatty acids levels increase during starvation, thus indicating further similarities in metabolic response to prolonged starvation between vertebrates and

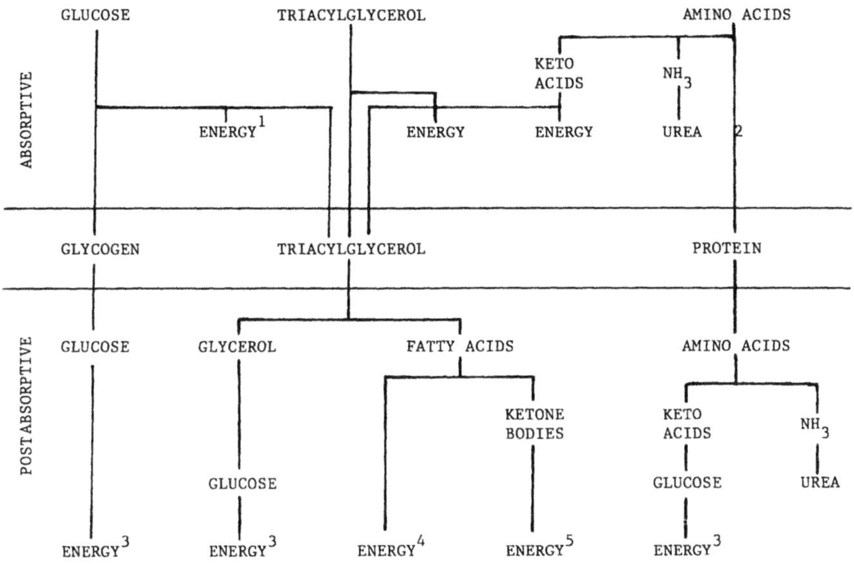

Fig. 7. Major routes of metabolic flow in vertebrates during absorptive and postabsorptive states (from accounts of Newsholme and Start, 1973; Vander *et al.*, 1980). [1]Glucose is the primary source of energy during the absorptive state. [2]Excess amino acids are *not* stored as protein. [3]Glucose is used primarily by the nervous system during the postabsorptive state. [4]Fatty acids are the primary source of energy during the postabsorptive state. [5]Ketone bodies may be used by the nervous system following prolonged starvation.

insects, or at least those insect species that consume lipid as their primary energy source during periods of high energy demand. Many studies have examined the fate of radiolabeled substrates injected or fed to insects that may be presumed to have been in the absorptive state (see reviews by Chippendale, 1978; Downer, 1978; Chen, 1978.) The results of these studies do not indicate any major differences in the metabolic pattern of insects in the absorptive state from that which has been described for vertebrates.

On the strength of the foregoing discussion it is reasonable to propose that the quantitative importance of particular metabolic pathways varies according to the nutritional state (absorptive, postabsorptive, starved) of the insect. The integration and regulation of the equivalent pathways in vertebrates is under endocrine control, with insulin, and, to a lesser extent, epinephrine, glucagon, and growth hormone serving to mediate precise metabolic control. It is likely that similar endocrine mechanisms are operative in insects, but, although several possibilities are revealed in the present volume, this crucial aspect of insect metabolism remains unexplained.

4.5. Circadian Rhythmicity

Even in the absence of external temporal signals, insects express a variety of biorhythms with periodicities that approximate 24 hours (Brady, 1973). Rhythmicity has been described for a number of biochemical parameters that warrant identification in the present discussion. A pronounced diel periodicity in hemolymph sugar levels occurs independently of behavioral rhythms in *Acheta domesticus* (Nowosielski and Patton, 1964) and *P. americana* (Hilliard and Butz, 1969), and a similar trehalose rhythm has been suggested in fourth instar larvae of *Aedes taeniorhynchus* (Nayar, 1969). Circulating levels of hydrocarbons show circadian rhythmicity in *P. americana,* with concentrations of heptacosadiene increasing during the forenoon while those of pentacosane and methylpentacosane decrease (Turner and Acree, 1967). Other metabolic rhythms in sensitivity to anesthesia and insecticide poisoning have also been reported (Brady, 1974). Such metabolic rhythms, together with well-established locomotive, feeding, and neurosecretory rhythms (Brady, 1973), accentuate the importance of circadian rhythmicity as a factor influencing metabolic flux in insects. The foregoing discussion serves also to emphasize the warning of Corbet (1966) that consideration of circadian rhythms is of great importance in the design and interpretation of entomological experiments.

5. Conclusion

The total energy consumed by an insect satisfies a variety of biological functions. The relative importance of these diverse functions varies at different times during the life of the insect and according to the conditions that exist at particular times. In order to meet the ever-changing demands of insect systems for energy, the bioenergetic metabolic pathways respond to a variety of physiological conditions, including those resulting from age, sex, developmental stage, temperature, excitation, and nutritional state. Particular physiological conditions are likely to affect the neuroendocrine system, which, in turn, influences the rate and degree of energy flow through the various metabolic pathways. The components of the neuroendocrine system that are involved in this precise and integrated regulation of organic metabolism and the manner in which these various components interact have not been elucidated. It is hoped that the present volume will stimulate interest in this important area of insect physiology/biochemistry and, in light of the extremely dynamic nature of insect metabolism, emphasize the need for precise definition of any system used in the investigation of bioenergetic metabolism.

ACKNOWLEDGMENTS. I am grateful to Dr. D.M. Lavigne for his critical reading of the manuscript and to Dr. V.L. Kallapur for permission to cite unpublished observations. Work from this laboratory was supported by a grant from the Natural Science and Engineering Research Council of Canada.

References

Agrell, I.P.S., and Lundquist, A.M., 1973, Physiological and biochemical changes during insect development, in: *The Physiology of Insecta* (M. Rockstein, ed.), pp. 159–247, Academic Press, New York.

Bell, W.J., and Bohm, M.K., 1975, Oosorption in insects, *Biol. Rev.* **50**:373.

Brady, J., 1973, Physiology of insect circadian rhythms, *Adv. Insect Physiol.* **10**:1.

Carlson, L.D., and Hsieh, A.C.L., 1970, *Control of Energy Exchange*, Macmillan, London.

Chen, P.S., 1971, *Biochemical Aspects of Insect Development*, S. Karger, Basel.

Chen, P.S., 1978, Protein synthesis in relation to cellular activation and deactivation, in: *Biochemistry of Insects* (M. Rockstein, ed.), pp. 145–203, Academic Press, New York.

Chippendale, G.M., 1978, The functions of carbohydrates in insect life processes, in: *Biochemistry of Insects* (M. Rockstein, ed.), pp. 1–55, Academic Press, New York.

Corbet, P.S., 1966, The role of rhythms in insect behaviour, in: *Insect Behaviour* (P.T. Haskell, ed.), pp. 13–28, *Symp. R. Entomol. Soc. Lond.* #3.

Downer, R.G.H., 1978, Functional role of lipids in insects, in: *Biochemistry of Insects* (M. Rockstein, ed.), pp. 57–92, Academic Press, New York.

Downer, R.G.H., 1979, Induction of hypertrehalosemia by excitation in *Periplaneta americana, J. Insect Physiol.* **25**:59.

Downer, R.G.H., 1980, Short-term hypertrehalosemia induced by octopamine in the American cockroach, *Periplaneta americana* L., in: *Insect Neurobiology and Pesticide Action (Neurotox 79)*, pp. 335–339, Society of Chemical Industry, London.

Downer, R.G.H., 1981, Fat body, in: *The American Cockroach* (W.J. Bell and K.G. Adiyodi, eds.), Chapman and Hall, London.

Downer, R.G.H., and Matthews, J.R., 1977, Production and utilisation of glucose in the American cockroach, *Periplaneta americana, J. Insect Physiol.* **23**:1429.

Downer, R.G.H., and Matthews, J.R., 1978, Trehalase activity of serum and whole haemolymph in the American cockroach, *Periplaneta americana* L., *Can. J. Zool.* **56**:2217.

Fast, P.G., 1970, Insect lipids, *Prog. Chem. Fats Other Lipids* **11**:181.

Gilbert, L.I., 1967, Lipid metabolism and function in insects, *Adv. Insect Physiol.* **4**:69.

Gole, J.W.D., and Downer, R.G.H., 1980, Elevation of adenosine 3′,5′-monophosphate by octopamine in fat body of the American cockroach, *Periplaneta americana* L., *Comp. Biochem. Physiol.* **64C**:223.

Heinrich, B., 1975, Thermoregulation in bumble bees. II. Energetics of warm up and free flight, *J. Comp. Physiol.* **96**:155.

Heinrich, B., and Bartholomew, G.A., 1971, An analysis of pre-flight warm-up in the sphinx moth, *Manduca sexta, J. Exp. Biol.* **55**:223.

Hill, L., Mordue W., and Highnam, K.C., 1966, The endocrine system, frontal ganglion, and feeding during maturation in the female desert locust, *J. Insect Physiol.* **12**:1197.

Hill, L., Izatt, M.E.G., Horne, J.A., and Bailey, E., 1972, Factors affecting concentrations of acetoacetate and D-3-hydroxybutyrate in hemolymph and tissues of the adult desert locust, *J. Insect Physiol.* **18**:1265.

Hilliard, S.D., and Butz, A., 1969, Daily fluctuations in the concentrations of total sugar and uric acid in the hemolymph of *Periplaneta americana*, *Ann. Entomol. Soc. Am.* **62**:71.

Jutsum, A.R., Agarwal, H.C., and Goldsworthy, G.J., 1975, Starvation and hemolymph lipids in *Locusta migratoria migratorioides* (R. and F.), *Acrida* **4**:47.

Kallapur, V.L., Downer, R.G.H., and George, J.C., 1980, Conversion of U-^{14}C-glucose to lipid in American cockroach, *Periplaneta americana* L., *Arch. Int. Physiol. Biochim.* **88**:363.

Katagiri, C., 1977, Localization of trehalase in the hemolymph of the American cockroach, *Periplaneta americana*, *Insect Biochem.* **7**:351.

Klekowski, R.Z., and Duncan, A., 1975, Physiological approach to ecological energetics, in: *Methods for Ecological Bioenergetics* (W. Grodzinski, R.Z. Klekowski, and A. Duncan, eds.), pp. 15–64, Blackwell, Oxford.

Lehninger, A.L., 1971, *Bioenergetics*, Benjamin, Menlo Park, California.

Lipke, H., Graves, B., and Leto, S., 1965, Polysaccharide and glycoprotein formation in the cockroach. II. Incorporation of D-glucose-^{14}C into bound carbohydrate, *J. Biol. Chem.* **24**:601.

MacArthur, R.H., and Wilson, E.O., 1967, *The Theory of Island Biogeography*, Princeton University Press, Princeton.

Matthews, J.R., Downer, R.G.H., and Morrison, P.E., 1976, α-Glucosidase activity in haemolymph of the American cockroach, *Periplaneta americana*, *J. Insect Physiol.* **22**:157.

May, M.L., 1979, Insect thermoregulation, *Ann. Rev. Entomol.* **24**:313.

Mwangi, R.W., and Goldsworthy, G.J., 1977, Diglyceride-transporting lipoproteins, in *Locusta*, *J. Comp. Physiol.* **114**:177.

Nathanson, J.A., 1979, Octopamine receptors, adenosine-3′,5′-monophosphate, and neural control of firefly flashing, *Science* **203**:68.

Nathanson, J.A., and Greengard, P., 1973, Octopamine-sensitive adenylate cyclase: Evidence for a biological role of octopamine in nervous tissue, *Science* **180**:308.

Nayar, J.K., 1969, The pubation rhythm in *Aedes taeniorhynchus*. V. Physiology of growth and endogenous diurnal rhythm of pupation, *Ann. Entomol. Soc. Am.* **62**:1079.

Newsholme, E.A., Crabtree, B., Higgins, S.J., Thornton, S.D., and Stalt, C., 1972, The activities of fructose diphosphatase in flight muscle from the bumble bee and the role of this enzyme in heat generation, *Biochem. J.* **128**:89.

Newsholme, E.A., and Start, C., 1973, *Regulation in Metabolism*, Wiley, London.

Nowosielski, J.W., and Patton, R.L., 1964, Daily fluctuations in the blood sugar concentrations of the house cricket, *Gryllus domesticus* L., *J. Insect Physiol.* **9**:401.

Prosser, C.L. 1973, *Comparative Animal Physiology*, 3rd ed., Saunders, Philadelphia.

Prus, T., 1975, Computation of energy budgets using all stages of *Tribolium*, in: *Methods for Ecological Bioenergetics* (W. Grodzinski, R.Z. Klekowski, and A. Duncan, eds.), pp. 263–272, Blackwell, Oxford.

Robertson, H.A., and Juorio, A.V., 1977, Octopamine and some related noncatecholic amines in invertebrate nervous systems, *Int. Rev. Neurobiol.* **19**:173.

Robertson, H.A., and Steele, J.E., 1973, Activation of insect nerve cord phosphorylase by octopamine and adenosine 3′,5′-monophosphate, *J. Neurochem.* **19**:1603.

Sacktor, B., 1970, Regulation of intermediary metabolism, with special reference to the control mechanisms in insect muscle, *Adv. Insect Physiol.* **7**:268.

Schmidt-Nielsen, K., 1980, *Animal Physiology: Adaptation and Environment*, Cambridge University Press, Cambridge.

Spring, J.H., Matthews, J.R., and Downer, R.G.H., 1977, Fate of glucose in haemolymph of the American cockroach, *Periplaneta americana*, *J. Insect Physiol.* **23**:525.

Turner, R.B., and Acree, F., Jr., 1967, The effect of photoperiod on the daily fluctuation of haemolymph hydrocarbons in the American cockroach, *J. Insect Physiol.* **13**:519.
Van Asperen, K., and Van Esch, I., 1956, The chemical composition of the haemolymph in *Periplaneta* with special reference to the mineral constituents, *Neth. J. Zool.* **11**:342.
Vander, A.J., Sherman, J.H., and Luciano, D.S., 1980, *Human Physiology: The Mechanisms of Body Function*, 3rd edition, McGraw-Hill, New York.
Van Handel, E., 1978, Trehalose turnover and the coexistence of trehalose and active trehalase in cockroach haemolymph, *J. Insect Physiol.* **24**:151.
Wlodawer, P., and Wiesniewska, A., 1965, Lipids in the haemolymph of wax moth larvae during starvation, *J. Insect Physiol.* **11**:11.

2

Hormonal Regulation of Substrate Transport and Metabolism

David J. Candy

1. Introduction

The rates of metabolic reactions in insect tissues vary considerably under different conditions and respond to physiological stresses that may arise during development, during environmental changes, or during flight. For example, substrate oxidation in insect flight muscle increases by more than one hundred times when an insect starts to fly (Sacktor, 1975). Such variations in metabolic rate require that the rate of substrate supply to the tissues should also respond to changes in demand. Substrate release and transport from storage sites is therefore regulated in accordance with changing physiological requirements, and part of this regulation is effected by hormones. Such hormonal regulation forms the topic of this review.

Insect hormones are of three chemical types. The first are the lipid hormones (juvenile hormones and ecdysones) responsible for controlling the growth and development of insects. These are rather slow-acting hormones. The second group are the peptide neurohormones which are synthesized by neurosecretory cells and are normally secreted into the circulation (hemolymph) via neurohemal organs. (Other neuropeptides such as proctolin are released locally in the region of their target cells, but because they do not appear to act via the circulation they will not be considered further here.) The third group are the amines such as octopamine, which are also released from nerve endings. These may variously

David J. Candy • Department of Biochemistry, University of Birmingham, Birmingham, United Kingdom.

function as neurotransmitters, as local neuromodulators, or through the general circulation as hormones. Only the latter function will be considered here.

Previous reviews relevant to this topic include those of Wyatt (1972), Goldsworthy and Mordue (1974), Bailey (1975), Steele (1976), Goldsworthy and Cheeseman (1978), Mordue and Stone (1979), and Stone and Mordue (1980).

1.1. Assessment of Physiological Relevance of Hormones

Our knowledge of the hormonal regulation of metabolism in insects has now reached a stage where it is important to evaluate critically the progress that has been made. One major problem that becomes apparent on reviewing this area is the variety of information obtained from different insect species using different experimental approaches. From this it is clear that different species of insect each have their own characteristic mechanisms of hormonal control, so that it is not possible to predict with any certainty that a particular hormonal effect observed in one species will also be found in another species. It is also clear that the degree of reliance to be placed on evidence obtained using different experimental procedures varies considerably. Because of this, it is useful to set out the type of information required to assess whether a proposed hormonal effect is of physiological significance. This information includes:

1. The chemical identity of the proposed hormone and its presence in an appropriate secreting gland.
2. The responses of target tissues to the hormone and the dose-dependence of such responses.
3. The release of the hormone in response to specific physiological stimuli.
4. The presence of the hormone in hemolymph *in vivo*, in a concentration sufficient to elicit a response in target tissues.

Other information valuable to an understanding of hormonal regulation includes:

5. The molecular mechanism of action of the hormone (e.g., effects on second messenger systems, membrane permeability, or synthesis of specific proteins.
6. The rate of turnover of hormone in the circulation (including a knowledge of rates of release into, and removal from, the hemolymph).
7. The action of possible analogues, agonists, and antagonists on the target cells (i.e., specificity of hormone receptors).

1.2. General Methods Used to Study Insect Hormones

A variety of different experimental approaches have been used to elucidate insect hormone action. These are not always identical to the techniques used to study vertebrate hormones, since insects are often able to survive surgical insults (removal of the head, etc.) which cannot be carried out on vertebrates. Some general approaches are described below.

1. Isolate and identify the hormone. Usually the most favorable source for isolation of a hormone is the endocrine or neurosecretory gland from which it originates. Such glands usually contain considerable stores of hormone at much higher concentrations relative to contaminating substances than other possible sources, such as whole insects. Occasionally other sources have been used. For example, blowfly heads can be obtained in large quantities by shaking frozen insects and then sieving the freed heads to separate them from other body parts. Various fractionation techniques such as chromatography and solvent extraction are then applied to extracts of the biological material. The fractions are tested for biological activity using a bioassay system based on the presumed physiological action of the hormone. The purity of the product must then be established and its structure determined by chemical techniques. If possible, the structure should be confirmed by chemical synthesis and demonstration of the biological activity of the synthetic product. At present, good evidence for the structures of insect hormones has been achieved for only a limited number of compounds—adipokinetic hormone (Section 4), the ecdysones, and the juvenile hormones.

Alternative identification procedures have been applied in certain cases. Thus, the presence of octopamine in insect tissues has been established by enzymic methylation with methyl-labeled *S*-adenosyl methionine as the methyl donor, followed by identification of the radioactive product (Section 5). This procedure has the advantage of considerable sensitivity, since the labeled product can be detected in very small amounts.

Immunological techniques have also been applied to insect systems. Thus, antibodies developed against purified mammalian hormones (insulin, glucagon, etc.) have been employed to recognize similar molecules in insects. However, although such techniques may be helpful in preliminary investigations, interpretation of the results is difficult because antibodies recognize and bind to only a part of the antigen molecule (the antigenic determinant). Thus, the extent of homology of insect immunoreactive substances with mammalian hormones is uncertain. It is also important with such techniques to establish the purity of the antibody preparation used, and this criterion does not seem to have been fulfilled in all cases. However, immunological techniques offer an important future prospect if specific antibodies to insect hormones (rather than their mammalian counterparts) become available.

2. The physiological effects of hormones can be studied in a number of different ways. One of the earliest techniques used was to remove a proposed hormone-producing gland from a living insect and to follow any subsequent physiological and biochemical changes that ensued. This has proved to be a valuable approach, but suffers from the disadvantage that often more than one hormone may be produced by a gland (e.g., the corpora cardiaca) so that multiple effects are produced owing to the absence of several hormones. Allatectomy (removal of the corpora allata) is a particularly useful technique because these glands are believed to secrete only juvenile hormones.

An alternative procedure used to study effects of hormones *in vivo* is to inject extracts of hormone-producing glands into recipient insects and to follow any resulting changes. This technique can then be extended to monitor partially-purified and purified hormonal extracts, so that problems associated with multiple hormonal action can be eliminated.

Because experiments carried out on insects *in vivo* can do no more than establish the overall effects of hormones, experiments with isolated tissues have been invaluable in discovering the target cells on which the hormone acts. This is particularly important where a hormone has different effects on different tissues, as occurs with juvenile hormone. *In vitro* incubations are also used to study hormonal effects on second messenger levels (e.g., of cAMP), effects of hormone analogues on the system, and effects of possible inhibitors (antagonists, metabolic inhibitors, ion chelators) on the hormonal response. These studies can give information about the mechanism of action of the hormone.

It is important for both *in vivo* and *in vitro* experiments to show that any effects produced by hormonal stimulation occur with physiological concentrations of hormone. In the past, a number of effects have been reported in which large amounts of hormone extracts were used, such that hormone concentrations were two or more orders of magnitude higher than that which could occur under normal physiological conditions. Such results must be regarded with suspicion until the effects can be confirmed with the use of nonpharmacological doses of hormone.

3. One of the more difficult aspects of hormone research has been to establish the presence of hormonal factors in the hemolymph, and to measure any concentration changes that may occur in response to physiological stimulation. Nevertheless, before the physiological relevance of a hormonal effect can be established, it is essential to show that the hormone is present in concentrations that are known to elicit a response in target tissues.

The principal difficulty encountered in designing suitable assay systems for hormones in body fluids is the low concentrations of active substances relative to other solutes such as amino acids, proteins, and lipids. In the case of peptide hormones, the only available method of detecting activity may be by a biological assay. It is then usually necessary to carry out a partial separation of hormone from other hemolymph components, while ensuring that no significant losses of active material occur.

One aim of this review is to examine to what extent these techniques have been applied successfully to the study of the hormonal regulation of substrate supply and metabolism. The individual hormones will be considered in turn, and the evidence for their physiological significance will be outlined.

2. Hyperglycemic Hormones

Factors which cause elevation of blood sugar in insects are termed hyperglycemic. In most insects the main blood sugar is trehalose, and it is the concentration of this disaccharide that is usually affected by hormones. In some studies specific assays for trehalose have been used to study hyperglycemic effects, but in other studies total blood sugars have been measured. Where effects are known to specifically raise trehalose concentrations, it is correct to refer to the factors as hypertrehalosemic, but here the original term hyperglycemic (Steele, 1961) will be used throughout. It should be noted that in some studies concentrations of glucose have also been found to be affected by hormones.

The first reports of hyperglycemic factors in insects were made by Steele (1961, 1963), who showed that injection of extracts of corpora cardiaca into cockroaches (*Periplaneta americana*) caused elevation of the hemolymph trehalose concentration. The increase in hemolymph trehalose was correlated with a decrease in glycogen content of the fat body and with activation of fat body phosphorylase. Significant effects could be obtained with amounts of extract equivalent to less than 0.01 of a pair of corpora cardiaca, suggesting that the effect could be of physiological significance. Since these original observations, hyperglycemic effects have been demonstrated in other insect species including other cockroaches (*Blaberus discoidalis*, Bowers and Friedman, 1963; *Leucophaea maderae*, Wiens and Gilbert, 1967a), blowflies (*Phormia regina*, Friedman, 1967; *Calliphora erythrocephala*, Normann and Duve, 1969), bees (*Apis mellifera*, Natalizi and Frontali, 1966), and stick insects (*Carausius morosus*, Dutrieu and Gourdoux, 1967). As well as effects on fat body, corpora cardiaca extracts promote glycogenolysis in cockroach nerve cord (Hart and Steele, 1973) and hindgut (Tolman and Steele, 1980).

Hyperglycemic effects cannot be demonstrated in all insects, however. Wiens and Gilbert (1967b) found that the phosphorylase of silkmoth (*Hyalophora cecropia*) fat body was not activated by corpora cardiaca extracts, nor did corpora cardiaca extracts from the silkmoth have effects on the cockroach. (The cockroach is generally used as a test insect since it responds to hyperglycemic factors from many other insect species.) Perhaps this is not too surprising, since many Lepidoptera use lipids rather than carbohydrates for energy metabolism. However, in another leipdopteran insect (the tobacco hornworm, *Manduca sexta*) Zeigler (1979) found that amounts of corpora cardiaca extract equivalent to 0.01 gland pairs was sufficient to activate fat body phosphorylase. In locusts (Goldsworthy, 1969), a hyperglycemic response was demonstrated in male insects six days after the final molt, but not in insects of other ages, despite the fact that extracts of corpora cardiaca from locusts of other ages have a hyperglycemic effect on cockroaches (Mordue and Goldsworthy, 1969). Again, it should be noted that locusts use more lipid than carbohydrate as substrates during long-term flight.

There is evidence that hyperglycemic hormone acts by activating the cAMP system and thus increases the proportion of phosphorylase in the active form. Steele (1964) showed activation of phosphorylase in cockroach fat body incubated with cAMP (at high concentrations) or with caffeine (which can inhibit hydrolysis of cAMP in tissues and thus elevates intracellular cAMP concentrations). More recently, Hanaoka and Takahashi (1977) found that activation of adenylate cyclase and elevation of cAMP levels in fat body occurred when crude extracts of corpora cardiaca were injected into cockroaches. Other results correlating hyperglycemic effects with cAMP have been obtained by Gäde (1977) and by Zeigler *et al.* (1979). However, in these experiments corpora cardiaca extracts were used as the hormone source, and such extracts are known to contain a number of physiologically active peptides including adipokinetic hormone, which also appears to activate adenylate cyclase of insect fat body (see Section 4). Despite the lack of conclusive evidence, it seems reasonable to assume that hyperglycemic factors in insects have a similar action to that of glucagon in mammals. That is, they activate adenylate cyclase, and elevate cAMP levels, thus activating protein kinase, which phosphorylates regulatory proteins. In mammalian liver, protein kinase catalyzes the phosphorylation of phosphorylase kinase, which in turn catalyzes the phosphorylation of phosphorylase b to the more active a form. A similar cascade could operate in insects, since a cAMP-dependent protein kinase is present in cockroach (*Periplaneta americana*) fat body (Hanaoka and Takahashi, 1978). Also, Ashida and Wyatt (1979) have demonstrated the presence of phosphorylase kinase in silkmoth fat body.

Whatever the detailed mechanism, activation of phosphorylase would lead to stimulation of glycogen breakdown to sugar phosphates, which in insect fat body are largely converted to trehalose. It is noteworthy that in mammalian liver, glucagon stimulates the synthesis of carbohydrate by gluconeogenesis (via cAMP-dependent protein kinase regulation of glycolytic and gluconeogenic enzymes—see Pilkis *et al.*, 1978), but that no such function has yet been shown for insect hyperglycemic hormone.

A number of attempts have been made to purify hyperglycemic hormones (Brown, 1965; Natalizi and Frontali, 1966; Natalizi *et al.*, 1970; Traina *et al.*, 1976; Holwerda *et al.*, 1977a,b). Although partial separation of hyperglycemic activity from other factors was achieved in these experiments, hyperglycemic hormone has not yet been obtained in a pure state from any insect, and it is not clear to what extent the structures of hyperglycemic factors from different insect species may prove to be chemically similar to one another. In several of these reports there is evidence that more than one fraction has hyperglycemic activity. A major problem of interpretation is that corpora cardiaca contain a number of chemically related peptides that at sufficiently high concentrations will activate other than their normal physiological targets. Thus, purified adipokinetic hormone from locusts is capable of causing hyperglycemic effects in cockroaches, although over ten times as much hormone is required to elicit the cockroach hyperglycemic response as to elicit an adipokinetic response in locusts (Jones *et al.*, 1977). Holwerda *et al.* (1977b) have separated two peptides from cockroach corpora cardiaca, one of which has predominantly hyperglycemic activity (as tested in cockroaches) and the other has adipokinetic activity (as tested in locusts). Neither appears to be identical to locust adipokinetic hormone. Jones and Mordue (quoted in Mordue and Stone, 1979) also report that cockroach hyperglycemic hormone is not identical to locust adipokinetic hormone, although their preparation is able to elicit an adipokinetic response in locusts.

Although the precise chemical structure of hyperglycemic hormone from cockroaches is not known, it is almost certainly a peptide, as early experiments on its susceptibility to proteolytic enzymes such as chymotrypsin showed. More recently, the anomalous behavior of hyperglycemic hormone on gel filtration (it is retarded beyond the void volume of the column) and its electrophoretic properties (Holwerda *et al.*, 1977b) suggest that it is an electrically neutral peptide. These properties are similar to those of adipokinetic hormone, which is the only hormone from corpora cardiaca whose structure is known. Taken together with the cross-reactivity of the two hormones in test systems, it seems likely that hyperglycemic hormone may well turn out to have structural similarities with adipokinetic hormone. There is general agreement that the major source of hyperglycemic hormone in all species studies is the corpora cardiaca, although smaller quantities of active factor may also be found in the corpora allata (Steele, 1969).

The physiological role of hyperglycemic hormone in insects is still not established. The fact that, at least in some species, a hyperglycemic response can be obtained with low concentrations of corpora cardiaca extract suggests that it does have a physiological role. However, the presence of hyperglycemic hormone in insect hemolymph has never been established, and until this is done the role of hyperglycemic factors as circulating hormones must remain in question. Evidence that hyperglycemic hormone has a physiological role during flight comes from work on the blowfly *Calliphora erythrocephala* by Vejbjerg and Normann (1974). They showed that normal flies are able to maintain a constant level of trehalose in their hemolymph during a 45-minute flight, despite the fact that trehalose is the main fuel used by the flight muscles of these insects. However, operated flies in which the nervous connections between the brain and the corpora cardiaca had been cut were unable to maintain blood trehalose during flight, so the concentrations dropped markedly to a point where the insects could no longer fly. When the corpora cardiaca of such flown operated insects were squeezed with fine forceps (to break up the tissue and presumably release some contents) there was a rapid recovery of trehalose levels towards normal, and the insects were able to resume flight. Since it is known that electrical stimulation of the connectives between brain and corpora cardiaca causes release of hyperglycemic factors in *Calliphora* (Normann and Duve, 1969) and in *Periplaneta* (Gersch *et al.*, 1970; Gersch, 1972), these experiments are reasonably interpreted in terms of release of hyperglycemic hormone during flight as a result of nervous stimulation of the corpora cardiaca.

Because of similarities in the action of hyperglycemic hormones and mammalian glucagon, a search has been made for a glucagon-like substance in insects. Thus, Tager *et al.* (1976) have demonstrated the presence of both glucagon-like and insulin-like immunoreactive material in extracts of the corpora cardiaca–corpora allata complex from the tobacco hornworm, *Manduca sexta*. The glucagon-like fractions caused depletion of fat body glycogen *in vivo* when tested in high concentration (equivalent to four gland complexes per test insect). High concentrations of mammalian glucagon also had a slight effect in decreasing fat body glycogen. Zeigler (1979) found that glucagon would activate fat body phosphorylase in *Manduca*, but only at concentrations some 10,000 times higher than those required to elicit an effect in mammals. Recently Kramer *et al.* (1980) have found that components in *Manduca* hemolymph also react with antibodies to mammalian insulin.

These findings are difficult to interpret. Firstly, because antibodies interact with only part of the antigen molecule (the antigenic determinant) it may be that only a part of the insect immunoreactive molecules bears any resemblance to part of the mammalian glucagon molecule. Chance homologies in sequence between functionally different molecules have been known to give misleading results when immunochemical studies are the only criteria used (Julliard *et al.*, 1980). Different mammalian neuropeptides can show immunochemical cross-reactivity (see Krieger and Liotta, 1979), which is not surprising since such peptides (e.g., β-endorphin and β-lipotropin) have amino acid sequences in common with each other. Secondly, the concentrations of both insect "glucagon-like" factors and of mammalian glucagon required to bring about effects in insects are much higher than could be considered to be physiological amounts. Whether the hyperglycemic factors in insects will prove to be identical to the glucagon-like substance, and to what extent they may be homologous with mammalian glucagon, awaits the isolation and sequence determination of these insect peptides.

3. Hypoglycemic Hormones

Evidence is now accumulating that neurosecretory cells in the brain are involved in regulation of carbohydrate metabolism in some insect species. Thomsen (1952) surgically removed the median neurosecretory cells from the brain of the blowfly *Calliphora erythrocephala* and found that glycogen accumulated in the fat bodies of such insects. Lea and Van Handel (1970) extended this finding to the mosquito *Aedes taeniorhynchus* and also showed that implantation of median neurosecretory cells into operated insects restricts glycogen synthesis. In *Locusta*, removal of the cerebral neurosecretory cells results in an accumulation of carbohydrate in fat body, flight muscles, and hemolymph (Goldsworthy, 1971). Effects on blood sugar were also noted by Normann (1975), who found that extirpation of the median neurosecretory cells of *Calliphora* caused a marked elevation of trehalose in the hemolymph within 30 hours of the operation. Injection of extracts of these neurosecretory cells brought about reversion of the hyperglycemia. Severing of the cardiac-recurrent nerve also caused hyperglycemia, and this result led to the suggestion that a hypoglycemic hormone synthesized in the median neurosecretory cells of the brain was passed along nerves to a neurohemal organ near the corpora cardiaca from which the hormone could be released into the aorta (Normann, 1975). Confirmation of this was obtained by Chen and Friedman (1977), who found that cardiatectomy of blowflies (*Phormia regina*) led to elevation of blood trehalose, as did cutting the nervous connections either anterior or posterior to the corpora cardiaca. Extracts of corpora cardiaca failed to bring about a decrease in trehalose levels in cardiactectomized insects, while brain extracts were active in this respect. It was therefore concluded that a hypoglycemic factor is synthesized in the brain, but is passed in nerves to beyond the corpora cardiaca before being released. This conclusion is in essential agreement with that of Normann. Duve (1978) noted that both trehalose and glucose levels of the hemolymph increased after extirpation of the median neurosecretory cells of *Calliphora*, and that the concentrations of both sugars were decreased after injection of extracts of these cells.

The mechanism of action of such proposed hypoglycemic factors is not known. Since carbohydrates of both hemolymph and fat body increase after removal of median neurosecretory cells, it is unlikely that the effect is solely on sugar conversion to glycogen. That is, hypoglycemic factors do not cause exactly opposite effects to those of hyperglycemic factors. It is possible that the overall balance of metabolism is affected. According to this suggestion, in the presence of hypoglycemic factors carbohydrate metabolism (i.e., uptake by tissues and conversion to other products) is stimulated, but in the absence of such factors the metabolism of other substrates (perhaps lipids) spares carbohydrate, which therefore accumulates. Alternatively, since most of the above experiments were carried out with flies fed *ad lib.*, perhaps feeding activity is increased in the absence of median neurosecretory cells, and excess carbohydrates are accumulated.

In summary, there is good evidence for blowflies in particular that the median neurosecretory cells are involved in restriction of blood sugar concentration, perhaps by secreting a hypoglycemic hormone into the hemolymph via a neurohemal organ. However, possible physiological roles for such factors are far from being understood. The factors have not yet been purified nor their structure determined, their presence in the circulation has not been established, their detailed mechanism of action remains obscure, and their response to different physiological conditions is not known. One obvious possibility (for which there is no experimental evidence) is that like insulin in mammals, hypoglycemic factors are released after intake of a carbohydrate meal to stimulate uptake and metabolism of blood sugars.

As with "insect glucagon," the similarity in action between the proposed hypoglycemic factors and insulin has led to a search for "insect insulin." Tager *et al.* (1976) fractionated extracts from the corpora cardiaca corpora allata complex of *Manduca sexta* and found a fraction that had immunoreactivity with antibodies developed against mammalian insulin. When this fraction was injected into test insects it brought about a decrease in the trehalose levels of the hemolymph, although large amounts were injected, and it is uncertain whether the effect could be of physiological relevance. More recently, the same group has been able to demonstrate the presence of immunoreactive insulin-like peptides in the hemolymph of *Manduca* (Kramer *et al.*, 1980). Insulin radioimmunoassays have been used to reveal the presence of insulin-like peptides in Hymenoptera (Ishay *et al.*, 1976; Duve and Thorpe, 1980). Mammalian insulin was found to reverse the hyperglycemia caused by removal of median neurosecretory cell from blowflies (Normann, 1975) and brought about hypoglycemia in honeybees (Ishay *et al.*, 1976; Bounias and Pachéco, 1979). However, high concentrations of insulin were used. As discussed for glucagon-like factors in insects (Section 2), the physiological significance of insulin-like molecules awaits further experimentation.

4. Adipokinetic Hormone

When locusts fly for more than a few minutes the lipid content of the hemolymph is elevated (Beenakkers, 1965). This increase in concentration is mainly in the diacylglycerol fraction (Mayer and Candy, 1967; Spencer and Candy, 1974), in keeping with the suggested role of diacylglycerol as the main transport form of lipid in a number of insect species and the form in which lipid is released from the triacylglycerol stores in the fat body (Chino and Gilbert, 1965; Tietz, 1967). Since the increase in hemolymph diacylglycerol during flight occurs despite a considerable increase in the rate of lipid oxidation (Weis-Fogh, 1952; Mayer and Candy, 1969a), it was proposed that there must be a factor which stimulates lipid mobilization during flight. A search revealed that such a factor was present in the locust corpora cardiaca (Mayer, 1968; Mayer and Candy, 1969b). Thus, extracts of corpora cardiaca from *Schistocerca gregaria* caused elevation of hemolymph diacylglycerol *in vivo*, with a time course similar to that found for the elevation of diacylglycerols during flight. Such extracts were also capable of promoting release of diacylglycerol from fat body incubated *in vitro*, suggesting that the fat body was the main target for the factor. A physiological role for the factor was suggested from the fact that hemolymph extracted from resting locusts contained no detectable amounts of the factor, whereas hemolymph taken from locusts flown for 90 minutes contained small but detectable amounts of an active component that had similar chromatographic properties to the factor from the corpora cardiaca. The active principle was found to be stable to boiling but inactivated by certain proteolytic enzymes. By analogy with lipid-mobilizing hormones in vertebrates, the factor is referred to as adipokinetic hormone.

These findings have been confirmed and extended by a number of research groups (e.g., Goldsworthy *et al.*, 1972a) working on both *Schistocerca gregaria* and *Locusta migratoria*. A considerable advance was made by Stone *et al.* (1976), who purified the hormone from locust corpora cardiaca and determined its chemical structure. This was the first complete identification of an insect peptide hormone. Adipokinetic hormone is a decapeptide (Fig. 1) in which the *N*-terminal amino acid is a pyro-

Adipokinetic hormone	pGlu – Leu – Asn – Phe – Thr – Pro – Asn – Trp – Gly – Thr – NH_2
Red pigment concentrating hormone	pGlu – Leu – Asn – Phe – Ser – Pro – Gly – Trp – NH_2

Fig. 1. Amino acid sequences of the adipokinetic hormone from locusts (Stone *et al.*, 1976) and the red-pigment-concentrating hormone from prawns (Fernlund, 1974).

glutamate residue, and the carboxyl group at the *C*-terminal threonine residue is present as an amide derivative. Since both of the end groups are uncharged, and since none of the side chains are ionizable at neutral pH, the whole molecule is neutral and does not migrate when subjected to electophoresis. It is soluble in methanol, and behaves anomalously during gel chromatography and chromatography on porous glass beads, being eluted later than would be predicted from its molecular weight (Stone *et al.*, 1976; Holwerda *et al.*, 1977b; Carlsen *et al.*, 1979). The structure of adipokinetic hormone has now been confirmed by chemical synthesis (Broomfield and Hardy, 1977), and the biological activity of the synthetic product has been shown to be identical to that of the naturally-occurring hormone (Stone *et al.*, 1978).

The structure of adipokinetic hormone is very similar to that of the red pigment-concentrating hormone (Fig. 1) isolated from prawns (Fernlund, 1974). Indeed, the crustacean hormone has an adipokinetic effect in locusts, although it is less potent than adipokinetic hormone, and adipokinetic hormone has a red pigment-concentrating effect in prawns, but again is less potent than the natural hormone (Mordue and Stone, 1976, 1977; Herman *et al.*, 1977; Stone *et al.*, 1978). Although the physiological effects in these two animals are very different, the speed of action (a few minutes) and origin in neurosecretory cells are similar, and it seems very likely that the two hormones have a common ancestral origin (see Mordue and Stone, 1979). It has also been noted that they are similar in size to some of the regulatory peptides of the mammalian hypothalamus and have in common with these the presence of an *N*-terminal pyroglutamate residue and an amide group on the *C*-terminal carboxyl group. The amino acid sequences of the invertebrate peptides, however, have little in common with the mammalian factors.

Recently Carlsen *et al.* (1979) have isolated a second peptide with adipokinetic activity from the locust corpora cardiaca. This ("AK-11") is present in smaller amounts than "AK-1" (the hormone of Stone *et al.*, 1976), but has a similar specific activity in both lipid-mobilizing effects in locusts and pigment-concentrating effects in crustacea. It is an octapeptide, and its amino acid composition is known but not the sequence (Carlsen *et al.*, 1979). Interestingly, Stone *et al.* (1978) have shown that an octapeptide "core" is required for analogues of adipokinetic hormone to have any biological activity. AK-11 is almost certainly a blocked peptide like AK-1, since it does not migrate during electrophoresis. The physiological significance of this second hormone is not yet known.

Goldsworthy *et al.* (1972a) showed that most, if not all, of the adipokinetic hormone content of the locust corpora cardiaca was present in the intrinsic (glandular) lobes rather than in the secretory lobes, which receive neurosecretory products from brain cells. It therefore seems likely that adipokinetic hormone is synthesized within the corpora cardiaca. Removal of the glandular lobes from locusts prevented the normal rise in hemolymph lipid in flown insects (Goldsworthy *et al.*, 1973). Similarly, severing the nervous connections between the brain and the corpora cardiaca prevented lipid elevation during flight, suggesting that nervous stimulation of the corpora cardiaca is essential for hormone release (Goldsworthy *et al.*, 1972b). The release of hormone can also be modulated by high concentrations of substrates in the hemolymph. Thus, injection of large amounts of trehalose (Cheeseman *et al.*, 1976) or lipid (Cheeseman and Goldsworthy, 1979) decreased the amount of adipokinetic hormone released in subsequently flown locusts. Electron microscopy of glandular lobes taken from flown and control locusts shows a difference in appearance indicative of secretory activity in the flown animals (Rademakers and Beenakkers, 1977). Adipokinetic hormone is assumed to be stored in neurosecretory granules, and in fact such granules have been partially purified from extracts of glandular lobes of locust corpora cardiaca (Stone and Mordue, 1979).

The presence of adipokinetic hormone in hemolymph of flown locusts (Mayer and Candy, 1969b) has been confirmed (Houben and Beenakkers, 1973; Cheeseman *et al.*, 1976; Cheeseman and Goldsworthy, 1979). The amount of adipokinetic hormone has been estimated to be of the order of 1 pmol/250 μl hemolymph for flown *Locusta* (Cheeseman and Goldsworthy, 1979), that is, an amount equivalent to about 0.5% of the total content of the corpora cardiaca (Stone *et al.*, 1976). It seems that there is an excess of reserve hormone in the corpora cardiaca, and this is in keeping with the situation in mammals, where there is a great excess of peptide hormones in the pituitary in relation to the amounts in the circulation. It is not, therefore, surprising that no difference can be found in the amounts of adipokinetic hormone in corpora cardiaca from flown and control insects (Mayer and Candy, 1969b).

Adipokinetic hormone may act on fat body through a cAMP-dependent system. Spencer and Candy (1976) incubated fat body with partially purified adipokinetic hormone in the presence of the phosphodiesterase inhibitor theophylline, and found that the cAMP concentrations in the tissue were elevated above control levels. Adipokinetic hormone also stimulated the incorporation of [^{3}H]adenine into cAMP. These effects could be produced by the equivalent of 0.01 gland pairs of hormone, that is, approximately physiological amounts. Gäde and Holwerda (1976) showed that there was an elevation of cAMP levels when extracts of corpora cardiaca (equivalent to 0.01 gland pairs) were injected into locusts, and Gäde and Beenakkers (1977) later found that it was the adipokinetic hormone content of these extracts that caused the effect. Supporting evidence for a role of cAMP in adipokinetic hormone action came from experiments in which dibutyryl cAMP or theophylline were found to mimic the effect of adipokinetic hormone in stimulating the release of diacylglycerol from locust fat body *in vitro* (Spencer and Candy, 1976).

By analogy with the action of mammalian adipokinetic hormones (such as catecholamines), it is suggested that locust adipokinetic hormone may bind to specific receptors on the fat body plasma membranes, and bring about activation of adenylate cyclase. The elevated concentrations of cAMP thus produced could then activate a protein kinase, which in turn could catalyze phosphorylation and activation of a regulatory enzyme, perhaps equivalent to the hormone-sensitive lipase of mammalian adipose tissue (see Robison *et al.*, 1971). The rate of breakdown of triacylglycerol to diacylglycerol plus nonesterified fatty acid would be enhanced, and the diacylglycerol would be passed to lipoprotein carriers in the hemolymph, while the nonesterified fatty acid is reesterified with glycerol phosphate (Candy *et al.*, 1976). (The latter part of the proposed process is in contrast to the process in mammalian adipose tissue, where complete hydrolysis of triacylglycerol occurs and nonesterified fatty acids plus glycerol are released into the blood.) Although this proposed mechanism for adipokinetic hormone action seems plausible, there is as yet little evidence for it, and efforts to find a hormone-sensitive lipase in locust fat body have proved unsuccessful (Spencer and Candy, 1976).

In addition to the physiological role of adipokinetic hormone in mobilizing lipid reserves from the fat body, the hormone may also regulate the balance of carbohydrate and lipid utilization during locust flight. Firstly, the release of adipokinetic hormone from the corpora cardiaca seems to be partly regulated by the concentration of trehalose in the hemolymph. As noted above, injection of trehalose into locusts to produce above-normal levels of trehalose in hemolymph inhibits adipokinetic hormone release and diacylglycerol mobilization in flown insects (Cheeseman *et al.*, 1976; Jutsum and Goldsworthy, 1977). Secondly, there is evidence that adipokinetic hormone acts directly on flight muscle to favor lipid oxidation. Thus, in perfused locust flight muscle, adipokinetic hormone stimulates the rate of oxidation of diacylglycerol (supplied as a lipoprotein complex) and inhibits the oxidation of glucose or trehalose (Robinson and Goldsworthy, 1977a). However, such effects cannot always be obtained (Candy, 1978). Surprisingly, extracts of corpora cardiaca were found to stimulate substrate oxidation by isolated flight muscle mitochondria (Robinson and Goldsworthy, 1977b), although the effect was only significant at the 5% level. If confirmed, this would be an unusual effect, since peptide hormones usually act through plasma membrane receptors. Extracts of corpora cardiaca have no effects on cAMP levels in flight muscle (Gäde and Holwerda, 1976), so adipokinetic hormone may operate through a different receptor system in muscle than in fat body.

Diacylglycerols are transported in insect hemolymph in the form of lipoproteins (see Chino, this volume), and in locusts there is a change in lipoprotein pattern in response to flight or injection of corpora cardiaca extracts. Mayer and Candy (1967) used cellulose-acetate electrophoresis to separate hemolymph proteins and found that in resting locusts (*Schistocerca*) diacylglycerols were associated with only one pair of lipoproteins (group A). Flown insects contained additional diacylglycerol in these lipoproteins and an extra pair of lipoproteins (group B) were found which also carried diacylglycerol. Mwangi and Goldsworthy (1977) used gel chromatography to separate proteins from *Locusta* hemolymph. In control insects, all the diacylglycerol was associated with one protein fraction, designated "A." Elevation of diacylglycerol concentrations in locusts by injection of corpora cardiaca extracts produced a new fraction, "A^+," which carried the extra diacylglycerol, and simultaneously a non-lipid-carrying protein "C" became depleted. It was proposed that during lipid mobilization some lipoprotein A combines with protein C plus diacylglycerol to form A^+, which has an increased lipid-carrying capacity. Interestingly, fledgling locusts two days after molting lacked protein C and also responded only weakly to adipokinetic hormone stimulation. The presence of protein C and the ability to respond to adipokinetic hormone both increase as the locusts develop until maximum response is reached after about eight days, which coincides with the attainment of maximum flight ability.

The changes in lipoproteins that occur during flight or after injection of adipokinetic hormone have been confirmed by Van der Horst *et al.* (1979). They extended the observations by using injections of disaccharides (trehalose or sucrose) to inhibit hormone release during flight. Such locusts showed no elevation of diacylglycerol concentrations in hemolymph, although the turnover of diacylglycerols was stimulated by flight. Injection of 0.05 gland equivalents of corpora cardiaca extract (equivalent to approximately ten times the estimated physiological amount of adipokinetic hormone) caused elevation of diacylglycerol concentrations and formation of A^+ lipoproteins. Van der Horst *et al.* (1979) suggested that the primary action of adipokinetic hormone might be on the association of hemolymph proteins, to produce A^+. However, it is difficult to envisage a molecular mechanism whereby adipokinetic hormone could effect such a change, and an alternative explanation would be that during flight in disaccharide-loaded insects there is a release of adipokinetic hormone at subnormal levels (which would be difficult to detect) to partly stimulate triacylglycerol breakdown. Thus, the type A lipoproteins may have sufficient capacity to accept diacylglycerol produced by fat body under these conditions, especially as the unloading of diacylglycerol from lipoproteins by muscles is increased during flight.

Although most of the information on adipokinetic hormone comes from work on locust species (*Schistocerca gregaria* and *Locusta migratoria*), there is some evidence for similar factors in other insect groups. Thus, Goldsworthy *et al.* (1972a) have reported an adipokinetic effect of corpora cardiaca extracts in the beetle *Tenebrio molitor*. Lepidoptera rely heavily on lipid oxidation to supply energy for flight (Beenakkers, 1969) and can therefore be assumed to regulate lipid transport. Dallman and Herman (1978) found that the elevation of hemolymph lipid which occurs in the monarch butterfly, *Danaus plexippus*, during flight could be mimicked by injection of extracts of head, brain, corpora cardiaca–corpora allata complex, or thoracic nerve. However, the physiological relevance of these effects was not established, as large quantities of the extracts were used. The active factor(s) appeared to be peptide in nature. Beenakkers *et al.* (1978) found that the tobacco hornworm, *Manduca sexta*, gave a hyperlipemic response to extracts of corpora cardiaca from hornworm, locust, or cockroach.

Hypolipemic effects of corpora cardiaca extracts have been reported for the cockroach *Periplaneta americana* (Downer, 1972; Downer and Steele, 1972). Thus, injection of extracts of either *Periplaneta* or *Locusta* corpora cardiaca into cockroaches caused a decrease in hemolymph lipid, which was accounted for mainly by a change in the diacylglycerol component. On the other hand, extracts of *Periplaneta* had a hyperlipemic effect in locusts. It was proposed that the fat bodies of the two species respond differently to similar factors from the corpora cardiaca. These differences may reflect different physiological requirements in the two species, and it was suggested that cockroaches may use the hormone to regulate lipid deposition during digestion of a meal.

Extracts of corpora cardiaca from the stick insect, *Carausius morosus*, have no effect on lipid or carbohydrate levels in the stick insect itself, yet have an adipokinetic effect on locusts and a hyperglycemic effect on cockroaches (Gäde, 1979). The physiological role of such factors in the stick insect is not known, but it may be that during evolution insects have developed a variety of different responses to a small range of chemically similar peptides. According to this idea, the factors from the corpora cardiaca of *Carausius* may have a different (unknown) physiological role in this insect and yet may be structurally similar to adipokinetic and hyperglycemic hormones.

5. Octopamine

Octopamine is a monophenolic amine derived from tyrosine. It is chemically related to the catecholamines (Fig. 2), being identical to noradrenaline except for the absence of one of the hydroxyl groups on the aromatic ring. Octopamine is widely distributed in nervous tissues of many vertebrates and invertebrates (for reviews see Robertson and Juorio, 1976; Axelrod and Saavedra, 1977; Evans, 1980). In insects and some other invertebrates there are larger amounts of octopamine than of catecholamines in nervous tissues (Robertson, 1976; Evans, 1978), so that octopamine may be of particular importance in these animals.

However, the exact roles of octopamine have yet to be established. Various suggestions have been made. Robertson and Carlson (1976) found that octopamine elicited light emission from firefly (*Photinus versicolor*) lanterns, and suggested that octopamine acts as a neurotransmitter in this system. In lobsters, specific octopaminergic neurons that may have a neurosecretory role have been identified in nerve trunks (Evans *et al.*, 1976b). It was proposed that octopamine could be released into the hemolymph from such neurons and might then affect various tissues such as heart, skeletal muscle, and blood cells (Evans *et al.*, 1976a; Battelle and

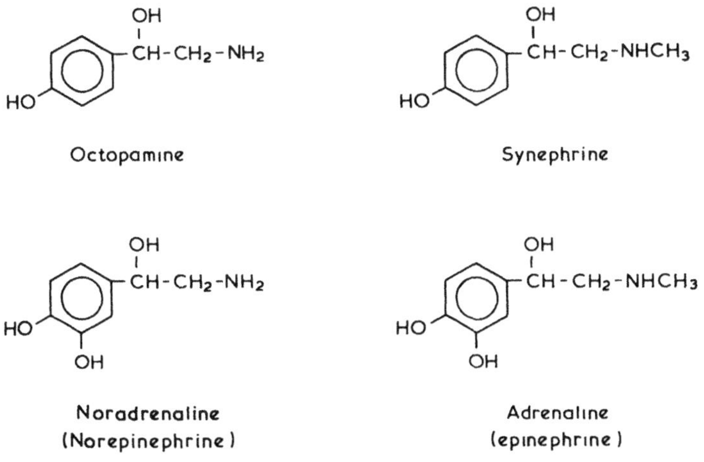

Fig. 2. Structures of octopamine and related amines.

Kravitz, 1978). Indeed, recently it has been shown that octopamine is present in lobster hemolymph, which would support the proposal of its possible role as a neurosecretory hormone (Livingstone *et al.*, 1980). However, it is not yet known whether octopamine concentrations in lobster hemolymph are altered in response to physiological stimuli.

In insects, the role of certain cells of the locust thoracic ganglion has been studied in some detail. These cells are known as the DUM (Dorsal Unpaired Median) cells, and they are probably octopaminergic. One of these cells ("DUMETi") innervates the extensor tibia muscles of the legs. Stimulation of the nerve containing the axon from this cell causes a slowing of the intrinsic rhythm of contraction of the leg, an effect which can also be produced by application of low concentrations of octopamine to the muscle (Hoyle, 1975; Evans and O'Shea, 1978). Octopamine is found in both cell body and axon of these cells (Evans and O'Shea, 1978) and it may be present within the neurosecretory dense core vesicles identified in these neurons (Hoyle, 1975; Hoyle *et al.*, 1980). It is presumed that octopamine can be released from nerve endings at the leg muscle, where it may have two effects. Firstly, by modulating the intrinsic rhythm of contraction of the muscle it may control muscular pumping of hemolymph to the extremities of the leg. Secondly, it potentiates the strength of contraction elicited by stimulation of the motor nerves (O'Shea and Evans, 1979), although octopamine does not itself act as a neuromuscular transmitter. It seems that in this system octopamine acts as a neuromodulator (local neurohormone), in some ways comparable to the role of noradrenaline in the mammalian sympathetic system.

Other DUM cells in the thoracic ganglia presumably innervate other muscles and perhaps other tissues in the locust thorax. Indeed, one such cell ("DUMDL") is known to supply the dorsal longitudinal muscles which are used during flight (Hoyle, 1978).

There is also evidence that octopamine may act as a circulating hormone in insects, and in particular may be of importance during locust flight. Hemolymph taken from resting locusts contains a low but significant concentration of D-octopamine (3×10^{-8} M), which rises during flight to a peak of about 17×10^{-8} M after 10 minutes (Fig. 3), and subsequently returns towards the resting value (Goosey and Candy, 1980a). This rapid rise in octopamine level resembles the increase in catecholamine concentration that can accompany exercise in mammals (Galbo *et al.*, 1975). It is proposed that the elevated octopamine levels could be of physiological significance in regulating flight muscles and perhaps fat body during the early period of flight. Thus, it is known that DL-octopamine increases the rate of oxidation of a number of substrates (glucose, trehalose, diacylglycerols) by perfused working thoracic muscles, and also increases the strength of muscular contractions (Candy,

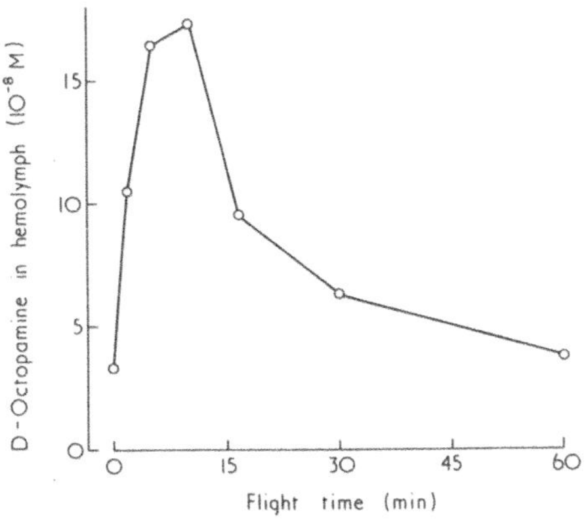

Fig. 3. Time course of changes in octopamine concentrations in locust hemolymph during flight (Goosey and Candy, 1980a).

1978). As with the locust leg muscle described above, octopamine modulates the effect caused by stimulation through motor nerves, and does not itself initiate contractions. Effects of DL-octopamine were significant at concentrations as low as 3×10^{-7} M. More recently (Goosey and Candy, 1980b), this effect on substrate oxidation has been found to be specific for the D isomer of octopamine (which is the isomer found in hemolymph), and significant effects were obtained with 1.5×10^{-7} M D-octopamine, which is approximately the concentration found in hemolymph during flight.

The origin of the octopamine found in hemolymph is not yet known. If some is released from octopamine-containing nerve terminals (such as those from DUM neurons) at the flight muscles it is possible that local concentrations of octopamine could be higher than is indicated from assays of hemolymph, and the effects on flight muscle would be correspondingly greater. Alternatively, the octopamine may be released from a more central source such as octopamine neurons in the central nervous system.

A number of practical difficulties are encountered in estimating the rather low concentrations of octopamine found in biological materials. The most commonly used method is that of Molinoff *et al.*, (1969) in which [^{14}C] or [^{3}H]methyl-*S*-adenosyl methionine is incubated with the sample in the presence of phenylethanolamine *N*-methyl transferase. The product of this reaction is labelled synephrine (a methylated derivative of octopamine—Fig. 2) which, in the original method, is separated from other components of the reaction mixture by solvent extraction. The radioactive incorporation into synephrine is then determined, and the identity of the product may be checked by chromatography. This method is sufficiently specific to allow good estimates of the amounts of octopamine in nervous tissues where the concentrations are reasonably high (e.g., Robertson, 1976; Evans, 1978). However, for hemolymph (and perhaps some other tissues) the concentrations of octopamine are much lower, so that interference by methylation of other compounds can give erroneously high results (Goosey and Candy, 1980a). This problem can be overcome by further purification of the labeled product by adding excess unlabeled carrier synephrine and then recrystallizing (three times) to constant specific radioactivity (Goosey and Candy, 1980a). Alternatively, the octopamine may first be separated from interfering compounds before carrying out the enzymic methylation (Livingstone *et al.*, 1980). The value of 6.5×10^{-8} M found by David and Lafon-Cazal (1979) for the concentration of octopamine in *Locusta* hemolymph is about two times the value obtained in hemolymph from resting *Schistocerca* (Goosey and Candy, 1980a). The value reported for *Locusta* may be too high, since it was obtained using only solvent extraction to isolate the product, whereas carrier crystallization was additionally used to ensure specificity for the assays on *Schistocerca* hemolymph.

Another biochemical effect of octopamine on substrate metabolism is a stimulation of glycogenolysis in cockroach (*Periplaneta americana*) nerve cord (Robertson and Steele, 1973). In this tissue it was found that 5×10^{-7} M DL-octopamine would activate phosphorylase and thus promote glycogen breakdown. Octopamine also stimulates glycogen breakdown to trehalose in cockroach fat body (Downer, 1979a,b; Gole and Downer, 1979). Significant effects were obtained with 10^{-6} M DL-octopamine, so this effect could be of physiological importance in stimulating carbohydrate transport from the fat body. It would be interesting to know whether cockroach hemolymph contains octopamine and whether levels change in response to different physiological factors.

There is now an accumulation of evidence that many, if not all, of the effects of octopamine on tissues are mediated through the elevation of intracellular cAMP. Nathanson and Greengard (1973) showed that the adenylate cyclase of cockroach ganglia is considerably stimulated by octopamine. Harmer and Horn (1977) confirmed this effect and showed that the enzyme was much more sensitive to D-octopamine than to the L isomer. Octopamine-sensitive adenylate cyclases have also been found in firefly (*Photinus pyrolis*) lanterns (Nathanson, 1979) and moth (*Mamestra configurata*) brains (Bodnaryk, 1979). Elevation of cAMP levels in tissues exposed to octopamine has been found in cockroach fat body (Gole and Downer, 1979) and lobster tissues (Battelle and Kravitz, 1978). In locust flight muscle the effects of octopamine are mimicked by theophylline (which inhibits breakdown of cAMP in tissues and hence tends to elevate cAMP levels) and by dibutyryl cAMP (Candy, 1978). Insect muscles are known to contain both adenylate cyclase (Arch and Newsholme, 1976) and cAMP-dependent protein kinase (Kuo *et al.*, 1971) and so should be capable of responding to stimuli by a cAMP-dependent system.

These results invite comparisons between the action of octopamine in insects and of catecholamines acting through β-receptors in mammals. However, insect octopamine receptors pharmacologically resemble mammalian α-receptors in that α-antagonists are frequently found to block octopamine effects and α-agonists elicit similar effects to those of octopamine. Usually β-antagonists and β-agonists have little effect on insect octopaminergic systems (Nathanson and Greengard, 1973; Harmer and Horn, 1977; Gole and Downer, 1979; Nathanson, 1979; O'Shea and Evans, 1979; Goosey and Candy, 1980b). It is noteworthy that mammalian catecholamines acting through α-receptors are believed to regulate cell processes by elevation of intracellular free calcium ion concentrations (see Exton, 1980), whereas β-receptors act through cAMP. There is at present no evidence for a role for calcium in octopamine actions in insects, and whether there is more than one type of octopamine receptor system remains an open question.

6. Juvenile Hormones and Ecdysones

The actions of peptide hormones and octopamine all involve rapid effects on target tissues, such that responses are observed within minutes of exposure to the hormone. Many of the actions of such hormones may be accounted for by effects on the activities of existing enzymes in the target cells, with little or no change in the total amounts of enzymes present.

Other insect hormones are mainly responsible for regulation of growth and development, processes which take days rather than minutes to accomplish. The two main groups of hormones (ecdysones and juvenile hormones) that bring about these changes are both lipid in nature, and their primary action seems to be on regulating the synthesis of specific proteins in the target cells. Unlike the peptide hormones and amine derivatives, which probably bind to receptors at the cell surface, it is likely that ecdysones and juvenile hormones have intracellular receptors that in some way affect the transcription process and hence the formation of specific messenger RNA molecules.

Substrate transport and metabolism is important during growth and development of insects, and ecdysones and juvenile hormones must therefore be involved, directly or indirectly, in the regulation of these processes. For example, it is clear that stored reserves must be mobilized during molting, since insects do not feed for a period immediately before or after ecdysis. However, little is known of the metabolic regulation of fuel transport during this time.

More is known of the role of juvenile hormones in controlling substrate transport and utilization in maturing adults, and one process which has been studied in some detail is the synthesis and transport of vitellogenins. Vitellogenins are female-specific proteins that are synthesized in the fat body of mature or maturing females (Telfer, 1954; Pan *et al.*, 1969). They are transported in the hemolymph to the developing ovaries and are there taken up into the growing oocytes (for reviews see Wyatt and Pan, 1978; Engelmann, 1979; Hagedorn and Kunkel, 1979). During egg production vitellogenins become the major proteins synthesized by the fat body, comprising 60% or more of the total secreted protein (e.g., Pan and Wyatt, 1976; Chen *et al.*, 1978). The vitellogenins are selectively taken up by the oocytes in a largely unmodified form and become the major protein ("vitellin") of the egg yolk. Vitellogenin and vitellin from the same insect species are usually immunologically indistinguishable from each other.

Vitellogenins supply most of the protein reserves of the eggs, and in many species also provide some of the lipid and carbohydrate reserves. Thus, locust vitellogenin is a glycolipoprotein containing about 10% lipid and 14% carbohydrate (mainly mannose) (Chen *et al.*, 1978). A similar lipid and carbohydrate content has been found in vitellogenins from other insect species including the cockroaches *Leucophaea* and *Blattella* (Koeppe and Ofengand, 1976; Kunkel and Pan, 1976) and the mosquito *Aedes* (Hagedorn and Judson, 1972).

The synthesis of vitellogenins in many insect species is controlled at least partly by juvenile hormones. Thus, allatectomy (removal of the corpora allata, which are the source of juvenile hormones) prevents vitellogenin synthesis in *Leucophaea* (Engelmann, 1969), *Periplaneta* (Bell, 1969), *Danaus* (Pan and Wyatt, 1976), and a number of other species. Vitellogenin synthesis can be restored in such allatectomized insects by injection or topical application of juvenile hormones (e.g., Bell, 1969; Engelmann, 1971). Juvenile hormone is also required for vitellogenin uptake by the oocytes (Bell, 1969). The effects of juvenile hormones are relatively slow to appear. For example, in *Leucophaea* vitellogenin synthesis is only apparent some 18 hours after treatment with juvenile hormone, and reaches a maximum after some 72 hours (Koeppe and Ofengand, 1976). Coordinated with the increase in vitellogenin production is a proliferation of endoplasmic reticulum in fat body, which occurs in response to juvenile hormone (della-Cioppa and Engelmann, 1980). This change is presumably required for the increased demand for synthesis of secreted proteins.

The time course of response to juvenile hormones is compatible with a period required for synthesis of messenger RNA coding for the synthesis of new protein. In keeping with this, stimulation of vitellogenin synthesis by juvenile hormone is blocked by actinomycin D (Engelmann, 1971) and α-amantin (Engelmann, 1976). Chen *et al.* (1976) used a wheat germ protein synthesis system to test for the ability of RNA from fat body of *Locusta* to stimulate vitellogenin synthesis. RNA from fat body of mature females stimulated synthesis of a protein that was identified as vitellogenin by immunoprecipitation. RNA from fat bodies of males or allatectomized females was ineffective. On the basis of this and other evidence it seems likely that juvenile hormones affect transcription of specific messenger RNA molecules, but the molecular mechanism by which this is achieved is not yet known.

The regulation of vitellogenin synthesis in the mosquito *Aedes aegypti* may differ in some respects from that of other insects studied (see Hagedorn and Kunkel, 1979). Both vitellogenin synthesis and ovary development in mosquitoes are triggered by a blood meal. It seems that such a meal may stimulate the ovaries to release β-ecdysone (Hagedorn *et al.*, 1975), and it is ecdysone rather than juvenile hormone that stimulates vitellogenin synthesis in the fat body. However, juvenile hormone is required prior to these events, otherwise the fat body is not competent to respond to the ecdysone (Flanagan and Hagedorn, 1977). β-Ecdysone stimulates vitellogenin synthesis by mosquito fat body *in vitro* if incubation times of 12 hours or more are used (Fallon *et al.*, 1974). However, the physiological relevance of ecdysone effects in mosquitoes has recently been disputed by Borovsky and Van Handel (1979), who found that only large concentrations of β-ecdysone, some 10,000 times physiological amounts, would initiate vitellogenin synthesis. They were also unable to induce vitellogenesis by transplanting ovaries from vitellogenic to nonvitellogenic females.

The ability of insect fat bodies to synthesize vitellogenins depends on factors in addition to juvenile hormone. Usually, fat bodies from male adults or early larval stages do not synthesize vitellogenins, despite the presence of juvenile hormone in the hemolymph of such insects. In some way the fat body of female adults is programmed to respond to juvenile hormones in a different way from the fat bodies of males and juveniles. Either neurosecretory factors or as yet undiscovered female-specific hormones could be involved in this programming. In some cases the concentration of juvenile hormone may be important, since vitellogenin synthesis can be induced prematurely in the last and second-to-last larval stages of the cockroach *Nauphoeta cinerea* by high doses of juvenile hormone, some 100 times that required to induce a response in adults (Lanzrein, 1974).

In adult locusts the accumulation of reserve materials, particularly lipids, is also partly controlled by juvenile hormones. There is a period of intense feeding activity during the first few days of adult life when much of the carbohydrate derived from food is converted to lipid in the fat body. At the end of this period the feeding activity decreases, as does the accumulation of lipid (Walker *et al.*, 1970), and these changes are correlated with the appearance of juvenile hormone in the hemolymph (Johnson and Hill, 1973). These results suggest that one function of juvenile hormone is to suppress lipid biosynthesis in the fat body. Such a response may be widespread in insects, since allatectomy results in excessive lipid accumulation in the fat bodies of *Calliphora erythrocephala* (Thomsen, 1952), *Periplaneta americana* (Vroman *et al.*, 1965), *Schistocerca gregaria* (Odhiambo, 1966; Walker and Bailey, 1971b), *Phormia regina* (Orr, 1964), and *Hyalophora cecropia* (Stephen and Gilbert, 1970). Similarly, the activity of lipogenic enzymes is higher in allatectomized locusts than in controls (Walker and Bailey, 1971a).

In female insects, the suppression of lipid synthesis by juvenile hormones can be seen as a part of the process of sexual maturation, in that lipids are thus made available for oocyte development. Thus, the lipids stored during the early part of adult life in female locusts are later transported to the ovary when juvenile hormone initiates maturation. In *Schistocerca*, there is an elevation of diacylglycerol concentration in transport lipoproteins of mature females that resembles that occurring during flight, when lipid transport is also activated (Mayer and Candy, 1967). Similarly, in the silkworm, *Philosamia cynthia*, there is a lipoprotein ("LP-1") which is responsible for transporting diacylglycerol from fat body to ovary during egg development (Chino *et al.*, 1977). At present, it is not known whether lipid transport to the ovary is regulated by adipokinetic hormone, juvenile hormone, or some as yet undiscovered factor. For the waxmoth, *Galleria mellonella*, it has been reported that a hormone from the ovaries may stimulate lipolytic activity in fat body (Dutkowski and Sarzala-Drabikowska, 1973). This would presumably have the effect of mobilizing lipid reserves from the fat body.

7. Other Hormones

Our present knowledge of hormonal regulation of insect metabolism is far from being complete, and it is likely that over the next few years additional hormones will be discovered. For example, it is known that insect nervous tissues contain a range of different substances that react with antibodies raised against mammalian hormones and neuropeptides. In addition to the glucagon-like and insulin-like substances discussed earlier, specific regions of insect nervous systems react with antibodies to mammalian vasopressin, vasotocin, neurophysin (Remy *et al.*, 1979; Remy and Giradie, 1980), enkephalin (Gros *et al.*, 1978), and gastrin (Kramer *et al.*, 1977). Such immunoreacting materials show specific distribution patterns in nervous tissues that are different for each antigen. At present, the functions of these putative peptides in insects are completely unknown, but by analogy with mammalian systems, they could function as neurosecretory hormones and/or as neurotransmitters. If so, there may be considerable potential for complex and sophisticated control of insect function.

One area of metabolism where possible hormonal regulation of metabolism appears to have been rather neglected is that of amino acid transport and utilization in insects. In some species, including the tsetse fly, *Glossina morsitans* (Bursell, 1963; this volume Chapter 5), the Colorado beetle, *Leptinotarsa decemlineata* (de Kort *et al.*, 1973; Mordue and de Kort, 1978), and the cockchafer, *Melolontha melolontha* (Crabtree and Newsholme, 1970), proline is the main fuel used by the thoracic muscles during flight. The product of proline oxidation is alanine, which is returned to the fat body via the hemolymph and is there converted back to proline (Fig. 4a). The additional carbons required for proline synthesis are derived by lipid breakdown to acetyl CoA (Bursell, 1977; Weeda *et al.*, 1980). By analogy with the proposed hormonal regulation of carbohydrate and lipid metabolism in other insect species, it seems possible that this process may also be hormonally controlled. One speculation could be that hormonal stimulation of lipid breakdown in fat body would provide an increased pool of acetyl CoA for conversion of alanine to proline. In some ways this proposal resembles that made for the action of adipokinetic hormone on glycerol metabolism in locusts, where the hormone stimulates glycerol incorporation into lipids. Thus, hydrolysis of triacylglycerol to diacylglycerol gives nonesterified fatty acids as a secondary product, and these are then reesterified with glycerol (Fig. 4b, Candy *et al.*, 1976). The analogy between the two systems is that alanine provides a carrier for fatty acid carbons in amino acid-metabolizing insects, and glycerol provides such a carrier in locusts.

Hormonal control of amino acid metabolism has been reported for *Schistocerca gregaria*, where cerebral neurosecretory cells are apparently required to maintain levels of transaminases in the fat body (Mordue and Goldsworthy, 1973). However, it is likely that these effects are rather slow-acting and may be more related to developmental changes than to the more rapid requirements of flight.

Effects of neurosecretory gland extracts, particularly from the corpora cardiaca, on stimulation of fat body respiration have also been described, e.g., for *Leucophaea maderae* (Wiens and Gilbert, 1967a; Müller

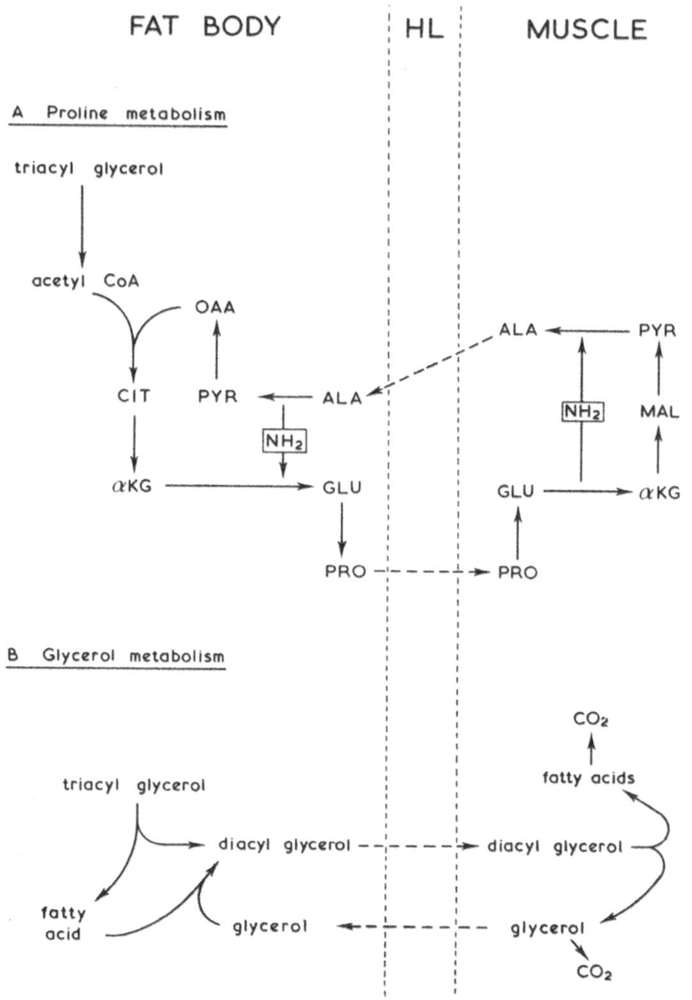

Fig. 4. Schemes for the roles of proline and glycerol in the transport of lipid from fat body to muscle during insect flight. A: Proline metabolism in the tsetse fly (Bursell, 1977). B: Glycerol metabolism in the locust (Candy *et al.*, 1976).

and Engelmann, 1968). However, since unfractionated extracts were used in these experiments it is not certain whether these are specific effects on respiration. Thus, stimulation of any energy-requiring process would lead to increased respiration as a secondary consequence of the primary action of the hormone. For example, stimulation of trehalose synthesis by hyperglycemic hormone would increase energy demand, as would triacylglycerol breakdown stimulated by adipokinetic hormone (since reesterification of the nonesterified fatty acid is energy-requiring). However, the long-term neuroendocrine regulation of mitochondrial respiratory function may be important, as discussed by Keeley (1978; this volume Chapter 8).

References

Arch, J.R.S., and Newsholme, E.A., 1976, Activities and some properties of adenylate cyclase and phosphodiesterase in muscle, liver and nervous tissues from vertebrates and invertebrates in relation to the control of the concentration of adenosine 3':5'-cyclic monophosphate, *Biochem. J.* **158**:603.

Ashida, M., and Wyatt, G.R., 1979, Properties and activation of phosphorylase kinase from silkmoth fat body, *Insect Biochem.* **9**:403.

Axelrod, J., and Saavedra, J.M., 1977, Octopamine, *Nature (London)* **265**:501.

Bailey, E., 1975, Biochemistry of insect flight. Part 2. Fuel supply, in: *Insect Biochemistry and Function* (D.J. Candy and B.A. Kilby, eds.), pp. 89–176, Chapman and Hall, London.

Battelle, B.A., and Kravitz, E.A., 1978, Targets of octopamine action in the lobster: Cyclic nucleotide changes and physiological effects in haemolymph, heart and exoskeletal muscle, *J. Pharmacol. Exp. Ther.* **205**:438.

Beenakkers, A.M.T., 1965, Transport of fatty acids in *Locusta migratoria* during sustained flight, *J. Insect Physiol.* **11**:879.

Beenakkers, A.M.T., 1969, Carbohydrate and fat as a fuel for insect flight. A comparative study, *J. Insect Physiol.* **15**:353.

Beenakkers, A.M.T., Van Der Horst, D.J., and Van Marrewijk, W.J.A., 1978, Regulation of release and metabolic function of the adipokinetic hormone in insects, in: *Comparative Endocrinology* (P.J. Gaillard and H.H. Boer, eds.), pp. 445–448, Elsevier/North Holland, Amsterdam.

Bell, W.J., 1969, Dual role of juvenile hormone in the control of yolk formation in *Periplaneta americana*, *J. Insect Physiol.* **15**:1279.

Bodnaryk, R.P., 1979, Identification of specific dopamine- and octopamine-sensitive adenylate cyclases in the brain of *Mamestra configurata*, *Insect Biochem.* **9**:155.

Borovsky, D., and Van Handel, E., 1979, Does ovarian ecdysone stimulate mosquitoes to synthesize vitellogenin?, *J. Insect Physiol.* **25**:861.

Bounias, M., and Pachéco, H., 1979, Susceptibility of honeybee glycaemia to injection of insulin and glucagon *in vivo*, *C.R. Acad. Sci.* **289**:201.

Bowers, W.S., and Friedman, S., 1963, Mobilization of fat body glycogen by an extract of corpus cardiacum, *Nature (London)* **198**:685.

Broomfield, C.E., and Hardy, P.M., 1977, The synthesis of locust adipokinetic hormone, *Tetrahedron Lett.* **25**:2201.

Brown, B.E., 1965, Pharmacologically active constituents of the cockroach corpora cardiaca: Resolution and some characteristics, *Gen. Comp. Endocrinol.* **5**:387.

Bursell, E., 1963, Aspects of the metabolism of amino acids in the tsetse fly, *Glossina* (Diptera), *J. Insect Physiol.* **9**:439.

Bursell, E., 1977, Synthesis of proline by fat body of the tsetse fly (*Glossina morsitans*): Metabolic pathways, *Insect Biochem.* **8**:427.

Candy, D.J., 1978, The regulation of locust flight muscle metabolism by octopamine and other compounds, *Insect Biochem.* **8**:177.

Candy, D.J., Hall, L.J., and Spencer, I.M., 1976, The metabolism of glycerol in the locust *Schistocerca gregaria* during flight, *J. Insect Physiol.* **22**:583.

Carlsen, J., Herman, W.S., Christensen, M., and Josefsson, L., 1979, Characterization of a second peptide with adipokinetic and red-pigment concentrating activity from the locust corpora cardiaca, *Insect Biochem.* **9**:497.

Cheeseman, P., and Goldsworthy, G.J., 1979, The release of adipokinetic hormone during flight and starvation in *Locusta*, *Gen. Comp. Endocrinol.* **37**:35.

Cheeseman, P., Jutsum, A.R., and Goldsworthy, G.J., 1976, Quantitative studies on the release of locust adipokinetic hormone, *Physiol. Entomol.* **1**:115.

Chen, A.C., and Friedman, S., 1977, Hormonal regulation of trehalose metabolism in the blowfly, *Phormia regina*: Interaction between hypertrehalosemic and hypotrehalosemic hormones, *J. Insect Physiol.* **23**:1223.

Chen, T.T., Couble, P., De Lucca, F.L., and Wyatt, G.R., 1976, Juvenile hormone control of vitellogenin synthesis in *Locusta migratoria*, in: *The Juvenile Hormones* (L.I. Gilbert, ed.), pp. 505–529, Plenum Press, New York.

Chen, T.T., Strahlendorf, P.W., and Wyatt, G.R., 1978, Vitellin and vitellogenin from locusts (*Locusta migratoria*): Properties and post-translational modification in the fat body, *J. Biol. Chem.* **253**:5325.

Chino, H., and Gilbert, L.I., 1965, Lipid release and transport in insects, *Biochim. Biophys. Acta* **98**:94.

Chino, H., Downer, R.G.H., and Takahashi, K., 1977, The role of diacylglycerol-carrying lipoprotein I in lipid transport during insect vitellogenesis, *Biochim. Biophys. Acta* **487**:508.

Crabtree, B., and Newsholme, E.A., 1970, The activities of proline dehydrogenase, glutamate dehydrogenase, aspartate-oxoglutarate aminotransferase and alanine-oxoglutarate aminotransferase in some insect flight muscles, *Biochem. J.* **117**:1019.

Dallman, S.H., and Herman, W.S., 1978, Hormonal regulation of haemolymph lipid concentration in the monarch butterfly, *Danaus plexippus*, *Gen. Comp. Endocrinol.* **36**:142.

David, J.C., and Lafon-Cazal, M., 1979, Octopamine distribution in the *Locusta migratoria* nervous and non-nervous systems, *Comp. Biochem. Physiol.* **64C**:161.

de Kort, C.A.D., Bartelink, A.K.M., and Schuurmans, R.R., 1973, The significance of L-proline for oxidative metabolism in the flight muscles of the colorado beetle, *Leptinotarsa decemlineata*, *Insect Biochem.* **3**:11.

della-Cioppa, G., and Engelmann, F., 1980, Juvenile hormone-stimulated proliferation of endoplasmic reticulum in fat body cells of a vitellogenic insect, *Leucophaea maderae* (Blattoria), *Biochem. Biophys. Res. Commun.* **93**:825.

Downer, R.G.H., 1972, Interspecificity of lipid-regulating factors from insect corpus cardiacum, *Can. J. Zool.* **50**:63.

Downer, R.G.H., 1979a, Trehalose production in isolated fat body of the American cockroach, *Periplaneta americana*, *Comp. Biochem. Physiol.* **62C**:31.

Downer, R.G.H., 1979b, Induction of hypertrehalosemia by excitation in *Periplaneta americana*, *J. Insect Physiol.* **25**:59.

Downer, R.G.H., and Steele, J.E., 1972, Hormonal stimulation of lipid transport in the American cockroach, *Periplaneta americana*, *Gen. Comp. Endocrinol.* **19**:259.

Dutkowski, A.B., and Sarzala-Drabikowska, M.G., 1973, Some aspects of regulation of fat body lipolytic activity in *Galleria mellonella*, *J. Insect Physiol.* **19**:1341.

Dutrieu, J., and Gourdoux, L., 1967, Le contrôle neuro-endocrine de la trehalosémie de *Carousius morosus*, *C.R. Acad. Sci.* **265D**:1067.

Duve, H., 1978, The presence of a hypoglycemic and hypotrehalocemic hormone in the neurosecretory system of the blowfly, *Calliphora erythrocephala*, *Gen. Comp. Endocrinol.* **36**:102.

Duve, H., and Thorpe, A., 1980, Isolation and localization of an insect insulin-like material: Immunological, biological and physical characteristics, *Gen. Comp. Endocrinol.* **40**:363.

Engelmann, F., 1969, Female specific proteins: Biosynthesis controlled by corpus allatum in *Leucophaea maderae*, *Science* **165**:407.
Engelmann, F., 1971, Juvenile hormone-controlled synthesis of female-specific protein in the cockroach *Leucophaea maderae*, *Arch. Biochem. Biophys.* **145**:439.
Engelmann, F., 1976, Induction of insect vitellogenin *in vivo* and *in vitro*, in: *The Juvenile Hormones* (L.I. Gilbert, ed.), pp. 470–485, Plenum Press, New York.
Engelmann, F., 1979, Insect vitellogenin: Identification, biosynthesis and role in vitellogenesis, *Adv. Insect Physiol.* **14**:49.
Evans, P.D., 1978, Octopamine distribution in the insect nervous system, *J. Neurochem.* **30**:1009.
Evans, P.D., 1980, Biogenic amines in the insect nervous system, *Adv. Insect Physiol.* **15**:317.
Evans, P.D., and O'Shea, M., 1978, The identification of an octopaminergic neurone and the modulation of a myogenic rhythm in the locust, *J. Exp. Biol.* **73**:235.
Evans, P.D., Kravitz, E.A., and Talamo, B.R., 1976a, Octopamine release at two points along lobster nerve trunks, *J. Physiol.* **262**:71.
Evans, P.D., Kravitz, E.A., Talamo, B.R., and Wallace, B.G., 1976b, The association of octopamine with specific neurones along lobster nerve trunks, *J. Physiol. (London)* **262**:51.
Exton, J.H., 1980, Mechanisms involved in α-adrenergic phenomena: Role of calcium ions in actions of catecholamines in liver and other tissues, *Am. J. Physiol.* **238**:E3.
Fallon, A.M., Hagedorn, H.H., Wyatt, G.R., and Laufer, H., 1974, Activation of vitellogenin synthesis in the mosquito *Aedes aegypti* by ecdysone, *J. Insect Physiol.* **20**:1815.
Fernlund, P., 1974, Structure of the red-pigment concentrating hormone of the shrimp, *Pandalus borealis*, *Biochim. Biophys. Acta* **371**:304.
Flanagan, T.R., and Hagedorn, H.H., 1977, Vitellogenin synthesis in the mosquito: The role of juvenile hormone in the development of responsiveness to ecdysone, *Physiol. Entomol.* **2**:173.
Friedman, S., 1967, The control of trehalose synthesis in the blowfly, *Phormia regina* Meig., *J. Insect Physiol.* **13**:397.
Gäde, G., 1977, Effect of corpus cardiacum extract on cyclic AMP concentration in the fat body of *Periplaneta americana*, *Zool. Jahrb. Abt. Allg. Zool. Physiol. Tiere* **81**:245.
Gäde, G., 1979, Adipokinetic and hyperglycaemic factor(s) in the corpora cardiaca/corpora allata complex of the stick insect, *Carausius morosus*. I. Initial characteristics, *Physiol. Entomol.* **4**:131.
Gäde, G., and Beenakkers, A.M.T., 1977, Adipokinetic hormone-induced lipid mobilization and cyclic AMP accumulation in the fat body of *Locusta migratoria* during development, *Gen. Comp. Endocrinol.* **32**:481.
Gäde, G., and Holwerda, D.A., 1976, Involvement of adenosine 3′:5′-cyclic monophosphate in lipid mobilization in *Locusta migratoria*, *Insect Biochem.* **6**:535.
Galbo, H., Holst, J.J., and Christensen, N.J., 1975, Glucagon and plasma catecholamine responses to graded and prolonged exercise in man, *J. Appl. Physiol.* **38**:70.
Gersch, M., 1973, Experimentelle Untersuchungen zum Freisetzungsmechanismus von Neurohormonen nach elektrischer Reizung der corpora cardiaca von *Periplaneta americana in vitro*, *J. Insect Physiol.* **18**:2425.
Gersch, M., Richter, K., Böhm, G.A., and Stürzebecher, J., 1970, Selektive Ausschüttung von Neurohormonen nach elektrischer Reizung der corpora cardiaca von *Periplaneta americana in vitro*, *J. Insect Physiol.* **16**:1991.
Goldsworthy, G.J., 1969, Hyperglycaemic factors from the corpus cardiacum of *Locusta migratoria*, *J. Insect Physiol.* **15**:2131.
Goldsworthy, G.J., 1971, The effects of removal of the cerebral neurosecretory cells on

haemolymph and tissue carbohydrates in *Locusta migratoria migratorioides*, *J. Endocrinol.* **50**:237.
Goldsworthy, G.J., and Cheeseman, P., 1978, Comparative aspects of the endocrine control of energy metabolism, in: *Comparative Endocrinology* (P.J. Gaillard and H.H. Boer, eds.), pp. 423–436, Elsevier/North Holland, Amsterdam.
Goldsworthy, G.J., and Mordue, W., 1974, Neurosecretory hormones in insects, *J. Endocrinol.* **60**:529.
Goldsworthy, G.J., Mordue, W., and Guthkelch, J., 1972a, Studies on insect adipokinetic hormones, *Gen. Comp. Endocrinol.* **18**:545.
Goldsworthy, G.J., Johnson, R.A., and Mordue, W., 1972b, *In vivo* studies on the release of hormones from the corpora cardiaca of locusts, *J. Comp. Physiol.* **79**:85.
Goldsworthy, G.J., Coupland, A.J., and Mordue, W., 1973, The effects of corpora cardiaca on tethered flight in the locust, *J. Comp. Physiol.* **82**:339.
Gole, J.W.D., and Downer, R.G.H., 1979, Elevation of adenosine 3′,5′-monophosphate by octopamine in fat body of the american cockroach, *Periplaneta americana* L., *Comp. Biochem. Physiol.* **64C**:223.
Goosey, M.W., and Candy, D.J., 1980a, The D-octopamine content of the haemolymph of the locust *Schistocerca americana gregaria* and its elevation during flight, *Insect Biochem.* **10**:393.
Goosey, M.W., and Candy, D.J., 1980b, Effects of D- and L-octopamine and of pharmacological agents on the metabolism of locust flight muscle, *Biochem. Soc. Trans.* **8**:532.
Gros, C., Lafon-Cazal, M., and Dray, F., 1978, Presence de substances immunoreactivement apparentées aux enképhalines chez un insect, *Locusta migratoria*, *C.R. Acad. Sci.* **287D**:647.
Hagedorn, H.H., and Judson, C.L., 1972, Purification and site of synthesis of *Aedes aegypti* yolk proteins, *J. Exp. Zool.* **182**:367.
Hagedorn, H.H., and Kunkel, J.G., 1979, Vitellogenin and vitellin in insects, *Annu. Rev. Entomol.* **24**:475.
Hagedorn, H.H., O'Connor, J.D., Fuchs, M.S., Sage, B., Schlaeger, D.A., and Bohm, M.K., 1975, The ovary as a source of α-ecdysone in an adult mosquito, *Proc. Natl. Acad. Sci. U.S.A.*, **72**:3255.
Hanaoka, K., and Takahashi, S.Y., 1977, Adenylate cyclase system and the hyperglycemic factor in the cockroach, *Periplaneta americana*, *Insect Biochem.* **7**:95.
Hanaoka, K., and Takahashi, S.Y., 1978, Endocrine control of carbohydrate metabolism, including the mechanism of action of the hyperglycemic hormone in insects, in: *Comparative Endocrinology* (P.J. Gaillard and H.H. Boer, eds.), pp. 455–458, Elsevier/North Holland, Amsterdam.
Harmer, A.J., and Horn, A.S., 1977, Octopamine-sensitive adenylate cyclase in cockroach brain: Effects of agonists, antagonists and guanylyl nucleotides, *Mol. Pharmac.* **13**:512.
Hart, D.E., and Steele, J.E., 1973, The glycogenolytic effect of the corpus cardiacum on the cockroach nerve cord, *J. Insect Physiol.* **19**:927.
Herman, W.S., Carlsen, J.B., Christensen, M., and Josefsson, L., 1977, Evidence for an adipokinetic function of the red pigment concentrating hormone activity present in the desert locust neuroendocrine system, *Biol. Bull., Woods Hole, Mass.* **153**:527.
Holwerda, D.A., Van Doorn, J., and Beenakkers, A.M.T., 1977a, Characterization of the adipokinetic and hyperglycaemic substances from the locust corpus cardiacum, *Insect Biochem.* **7**:151.
Holwerda, D.A., Weeda, E., and Van Doorn, J.M., 1977b, Separation of the hyperglycaemic and adipokinetic factors from the cockroach corpus cardiacum, *Insect Biochem.* **7**:477.
Houben, N.M.D., and Beenakkers, A.M.T., 1973, Regulation of diglyceride release from fat body of the locust during flight, *J. Endocr.* **57**:liv.

Hoyle, G., 1975, Evidence that insect dorsal unpaired median (DUM) neurons are octopaminergic, *J. Exp. Zool.* **193:**425.

Hoyle, G., 1978, The dorsal unpaired median neurons of the locust metathoracic ganglion, *J. Neurobiol.* **9:**43.

Hoyle, G., Colquoun, W., and Williams, M., 1980, Fine structure of an octopaminergic neuron and its terminals, *J. Neurobiol.* **11:**103.

Ishay, J., Gitter, S., Galun, R., Doron, M., and Laron, Z., 1976, The presence of insulin in, and some effects of exogenous insulin on hymenopteran tissues and body fluids, *Comp. Biochem. Physiol.* **54A:**203.

Johnson, R.A., and Hill, L., 1973, Quantitative studies on the activity of the corpora allata in adult male *Locusta* and *Schistocerca*, *J. Insect Physiol.* **19:**2459.

Jones, J., Stone, J.V., and Mordue, W., 1977, The hyperglycaemic activity of locust adipokinetic hormone, *Physiol. Entomol.* **2:**185.

Julliard, J.H., Shibasaki, T., Ling, N., and Guillemin, R., 1980, High-molecular-weight immunoreactive β-endorphin in extracts of human placenta is a fragment of immunoglobulin G, *Science* **208:**183.

Jutsum, A.R., and Goldsworthy, G.J., 1977, The role of the glandular lobes of the corpora cardiaca during flight in *Locusta*, *Physiol. Entomol.* **2:**125.

Keeley, L.L., 1978, Endocrine regulation of fat body development and function, *Annu. Rev. Entomol.* **23:**329.

Koeppe, J., and Ofengand, J., 1976, Juvenile hormone-induced biosynthesis of vitellogenin in *Leucophaea maderae*, *Arch. Biochem. Biophys.* **173:**100.

Kramer, K.J., Speirs, R.D., and Childs, C.N., 1977, Immunochemical evidence for a gastrin-like peptide in insect neuroendocrine system. *Gen. Comp. Endocrinol.* **32:**423.

Kramer, K.J., Tager, H.S., and Childs, C.N., 1980, Insulin-like and glucagon-like peptides in insect haemolymph, *Insect Biochem.* **10:**179.

Krieger, D.T., and Liotta, A.S., 1979, Pituitary hormones in brain: Where, how and why?, *Science* **205:**366.

Kunkel, J.G., and Pan, M.L., 1976, Selectivity of yolk protein uptake: Comparisons of vitellogenins of two insects, *J. Insect Physiol.* **22:**809.

Kuo, J.F., Wyatt, G.R., and Greengard, P., 1971, Cyclic nucleotide-dependent protein kinases. IX. Partial purification and some properties of cyclic GMP-dependent and cyclic AMP-dependent protein kinases from various species of Arthropoda, *J. Biol. Chem.* **246:**7159.

Lanzrein, B., 1974, Influence of a juvenile hormone analogue on vitellogenin synthesis and oogenesis in larvae of *Nauphoeta cinerea*, *J. Insect Physiol.* **20:**1871.

Lea, A.O., and Van Handel, E., 1970, Suppression of glycogen synthesis in the mosquito by a hormone from the medial neurosecretory cells, *J. Insect Physiol.* **16:**319.

Livingstone, M.S., Harris-Warrick, R.M., and Kravitz, E.A., 1980, Serotonin and octopamine produce opposite postures in lobsters, *Science* **208:**76.

Mayer, R.J., 1968, Flight metabolism in the desert locust, *Schistocerca gregaria*, Ph.D. Thesis, University of Birmingham.

Mayer, R.J., and Candy, D.J., 1967, Changes in haemolymph lipoproteins during locust flight, *Nature (London)* **215:**987.

Mayer, R.J., and Candy, D.J., 1969a, Changes in energy reserves during flight of the desert locust, *Schistocerca gregaria*, *Comp. Biochem. Physiol.* **31:**409.

Mayer, R.J., and Candy, D.J., 1969b, Control of haemolymph lipid concentration during locust flight: An adipokinetic hormone from the corpora cardiaca, *J. Insect Physiol.* **15:**611.

Molinoff, P.B., Landsberg, L., and Axelrod, J., 1969, An enzymic assay for octopamine and other β-hydroxylated phenylethylamines, *J. Pharmacol Exp. Ther.* **170:**253.

Mordue, W., and de Kort, C.A.D., 1978, Energy substrates for flight in the colorado beetle, *Leptinotarsa decemlineata*, *J. Insect Physiol.* **24**:221.

Mordue, W., and Goldsworthy, G.J., 1969, The physiological effects of corpora cardiaca extracts in locusts, *Gen. Comp. Endocrinol.* **12**:360.

Mordue, W., and Goldsworthy, G.J., 1973, Transaminase levels and uric acid production in adult locusts, *Insect Biochem.* **3**:419.

Mordue, W., and Stone, J.V., 1976, Comparison of the biological activities of an insect and a crustacean neurohormone that are structurally similar, *Nature* (*London*) **264**:287.

Mordue, W., and Stone, J.V., 1977, Relative potencies of locust adipokinetic hormone and prawn red pigment concentrating hormone in insect and crustacean systems, *Gen. Comp. Endocrinol.* **33**:103.

Mordue, W., and Stone, J.V., 1979, Insect hormones, in: *Hormones and Evolution* (E.J.W. Barrington, ed.), pp. 215–271, Academic Press, New York.

Müller, H.P., and Engelmann, F., 1968, Studies on the endocrine control of metabolism in *Leucophaea maderae* (Blattoria). 2. The effect of the corpora cardiaca on fat body respiration, *Gen. Comp. Endocrinol.* **11**:43.

Mwangi, R.W., and Goldsworthy, G.J., 1977, Diglyceride-transporting lipoproteins in *Locusta*, *J. Comp. Physiol.* **114**:177.

Natalizi, G.M., and Frontali, N., 1966, Purification of insect hyperglycaemic and heart accelerating hormones, *J. Insect Physiol.* **12**:1279.

Natalizi, G.M., Pansa, M.C., D'Ajello, V., Casaglia, O., Bettini, S., and Frontali, N., 1970, Physiologically active factors from corpora cardiaca of *Periplaneta americana*, *J. Insect Physiol.* **16**:1827.

Nathanson, J.A., 1979, Octopamine receptors, adenosine 3′-5′-monophosphate, and neural control of firefly flashing, *Science* **203**:65.

Nathanson, J.A., and Greengard, P., 1973, Octopamine-sensitive adenylate cyclase: Evidence for a biological role of octopamine in nervous tissues, *Science* **180**:308.

Normann, T.C., 1975, Neurosecretory cells in insect brain and production of hypoglycaemic hormone, *Nature* (*London*) **254**:259.

Normann, T.C., and Duve, H., 1969, Experimentally induced release of a neurohormone influencing haemolymph trehalose level in *Calliphora erythrocephala* (Diptera), *Gen. Comp. Endocrinol.* **12**:449.

Odhiambo, T.R., 1966, The metabolic effects of the corpus allatum hormone in the male desert locust. 1. Lipid metabolism, *J. Exp. Biol.* **45**:45.

Orr, C.W.M., 1964, The influence of nutritional and hormonal factors on the chemistry of the fat body, blood and ovaries of the blowfly *Phormia regina* Meig., *J. Insect Physiol.* **10**:103.

O'Shea, M., and Evans, P.D., 1979, Potentiation of neuromuscular transmission by an octopaminergic neurone in the locust, *J. Exp. Biol.* **79**:169.

Pan, M.L., and Wyatt, G.R., 1976, Control of vitellogenin synthesis in the monarch butterfly by juvenile hormone, *Dev. Biol.* **54**:127.

Pan, M.L., Bell, W.J., and Telfer, W.H., 1969, Vitellogenic blood protein synthesis by insect fat body, *Science* **165**:393.

Pilkis, S.J., Park, C.R., and Claus, T.H., 1978, Hormonal control of hepatic gluconeogenesis, *Vitam. Horm. N.Y.* **36**:383.

Rademakers, L.H.P.M., and Beenakkers, A.M.T., 1977, Changes in the secretory activity of the glandular lobe of the corpus cardiacum of *Locusta migratoria* induced by flight. A quantitative electron microscope study, *Cell Tissue Res.* **180**:155.

Rémy, C., and Giradie, J., 1980, Anatomical organization of two vasopressin-neurophysin-like neurosecretory cells throughout the central nervous system of the migratory locust, *Gen. Comp. Endocrinol.* **40**:27.

Rémy, C., Giradie, J., and Dubois, M.P., 1979, Vertebrate neuropeptide-like substances in the suboesophageal ganglion of two insects: *Locusta migratoria* R. and F. (Orthoptera) and *Bombyx mori* L. (Lepidoptera): Immunocytological investigation, *Gen. Comp. Endocrinol.* **37**:93.

Robertson, H.A., 1976, Octopamine, dopamine and noradrenaline content of the brain of the locust, *Schistocerca gregaria*, *Experienta* **32**:552.

Robertson, H.A., and Carlson, A.D., 1976, Octopamine: Presence in firefly lantern suggests a transmitter role, *J. Exp. Zool.* **195**:159.

Robertson, H.A., and Juorio, A.V., 1976, Octopamine and related non-catecholic amines in invertebrate nervous systems, *Int. Rev. Neurobiol.* **19**:173.

Robertson, H.A., and Steele, J.E., 1973, Effect of monophenolic amines on glycogen metabolism in the nerve cord of the american cockroach, *Periplaneta americana*, *Insect Biochem.* **3**:53.

Robinson, N.L., and Goldsworthy, G.J., 1977a, Adipokinetic hormone and the regulation of carbohydrate and lipid metabolism in a working flight muscle preparation, *J. Insect. Physiol.* **23**:9.

Robinson, N.L., and Goldsworthy, G.J., 1977b, A possible site of action for adipokinetic hormone on the flight muscle of locusts, *J. Insect Physiol.* **23**:153.

Robison, G.A., Butcher, R.W., and Sutherland, E.W., 1971, *Cyclic AMP*, Academic Press, New York.

Sacktor, B., 1975, Biochemistry of insect flight. Part 1. Utilization of fuels by muscle, in: *Insect Biochemistry and Function* (D.J. Candy and B.A. Kilby, eds.), pp. 1–88, Chapman and Hall, London.

Spencer, I.M., and Candy, D.J., 1974, The effect of flight on the concentrations and composition of haemolymph diacyl glycerols in the desert locust, *Biochem. Soc. Trans.* **2**:1093.

Spencer, I.M., and Candy, D.J., 1976, Hormonal control of diacyl glycerol mobilization from fat body of the desert locust, *Schistocerca gregaria*, *Insect Biochem.* **6**:289.

Steele, J.E., 1961, Occurrence of a hyperglycaemic factor in the corpus cardiacum of an insect, *Nature (London)* **192**:680.

Steele, J.E., 1963, The site of action of insect hyperglycaemic hormone, *Gen. Comp. Endocr.* **3**:46.

Steele, J.E., 1964, The activation of phosphorylase in an insect by cyclic AMP and other agents, *Am. Zool.* **4**:328.

Steele, J.E., 1969, The hyperglycaemic activity of the corpus allatum in an insect, *J. Insect Physiol.* **15**:421.

Steele, J.E., 1976, Hormonal control of metabolism in insects, *Adv. Insect Physiol.* **12**:239.

Stephen, W.F., and Gilbert, L.I., 1970, Alterations in fatty acid composition during the metamorphosis of *Hyalophora cecropia*: Correlations with juvenile hormone titre, *J. Insect Physiol.* **16**:851.

Stone, J.V., and Mordue, W., 1979, Isolation of granules containing adipokinetic hormone from locust corpora cardiaca by differential centrifugation, *Gen. Comp. Endocrinol.* **39**:543.

Stone, J.V., and Mordue, W., 1980, Isolation of insect neuropeptides, *Insect Biochem.* **10**:229.

Stone, J.V., Mordue, W., Batley, K.E., and Morris, H.R., 1976, Structure of locust adipokinetic hormone, a neurohormone that regulates lipid utilization during flight, *Nature (London)* **263**:207.

Stone, J.V., Mordue, W., Broomfield, C.E., and Hardy, P.M., 1978, Structure–activity relationships for the lipid-mobilizing action of adipokinetic hormone. Synthesis and activity of a series of hormone analogues, *Eur. J. Biochem.* **89**:195.

Tager, H.S., Markese, J., Kramer, K.J., Speirs, R.D., and Childs. C.N., 1976, Glucagon-like and insulin-like hormones of the insect neurosecretory system, *Biochem. J.* **156:**515.

Telfer, J.R., 1954, Immunological studies of insect metamorphosis. II. The role of a sex-limited blood protein in egg formation by the cecropia silkworm, *J. Gen. Physiol.* **37:**539.

Thomsen, E., 1952, Functional significance of the neurosecretory brain cells and the corpus cardiacum in the female blowfly, *Calliphora erythrocephala Meig., J. Exp. Biol.* **29:**137.

Tietz, A., 1967, Fat transport in the locust: The role of diglycerides, *Eur. J. Biochem.* **2:**236.

Tolman, J.H., and Steele, J.E., 1980, The control of glycogen metabolism in the cockroach hindgut: The effect of the corpora cardiaca-corpora allata system, *Comp. Biochem. Physiol.* **66B:**59.

Traina, M.E., Bellino, M., Serpietri, L., Massa, A., and Frontali, N., 1976, Heart accelerating peptides from cockroach corpora cardiaca, *J. Insect Physiol.* **22:**323.

Van der Horst, D.J., Van Doorn, J.M., and Beenakkers, A.M.T., 1979, Effects of the adipokinetic hormone on the release and turnover of haemolymph diglycerides and on the formation of the diglyceride-transporting lipoprotein system during locust flight, *Insect Biochem.* **9:**627.

Vejbjerg, K., and Normann, T.C., 1974, Secretion of hyperglycaemic hormone from the corpus cardiacum of flying blowflies, *Calliphora erythrocephala, J. Insect Physiol.* **20:**1189.

Vroman, H.E., Kaplanis, J.N., and Robbins, W.E., 1965, Effect of allatectomy on lipid biosynthesis and turnover in the female american cockroach, *Periplaneta americana* (L), *J. Insect Physiol.* **11:**897.

Walker, P.R., and Bailey, E., 1971a, Effect of allatectomy on the growth of the male desert locust during adult development, *J. Insect Physiol.* **17:**1125.

Walker, P.R., and Bailey, E., 1971b, Effect of allatectomy on fat body lipogenic enzymes of the male desert locust during adult development, *J. Insect Physiol.* **17:**1359.

Walker, P.R., Hill, L., and Bailey, E., 1970, Feeding activity, respiration, and lipid and carbohydrate content of the male desert locust during adult development, *J. Insect Physiol.* **16:**1001.

Weeda, E., Koopmanschap, A.B., de Kort, C.A.D., and Beenakkers, A.M.T., 1980, Proline synthesis in fat body of *Leptinotarsa decemlineata, Insect Biochem.* **10:**631.

Weis-Fogh, T., 1962, Fat combustion and metabolic rate of flying locusts (*Schistocerca gregaria* Forsk.), *Philos. Trans. R. Soc. London, Ser. B* **237:**1.

Wiens, A.W., and Gilbert, L.I., 1967a, Regulation of carbohydrate mobilization and utilization in *Leucophaea maderae, J. Insect Physiol.* **13:**779.

Wiens, A.W., and Gilbert, L.I., 1967b, The phosphorylase system of the silkworm, *Hyalophora cecropia, Comp. Biochem. Physiol.* **21:**145.

Wyatt, G.R., 1972, Insect hormones, in: *Biochemical Actions of Hormones*, Volume 2 (G. Litwack, ed.), pp. 385–490, Academic Press, New York.

Wyatt, G.R., and Pan, M.L., 1978, Insect plasma proteins, *Annu. Rev. Biochem.* **47:**779.

Ziegler, R., 1979, Hyperglycaemic factor from the corpora cardiaca of *Manduca sexta* (Lepidoptera: Sphingidae), *Gen. Comp. Endocrinol.* **39:**350.

Ziegler, R., Ashida, M., Fallon, A.M., Wimer, L.T., Wyatt, S.S., and Wyatt, G.R., 1979, Regulation of glycogen phosphorylase in fat body of Cecropia silkmoth pupae, *J. Comp. Physiol.* **131:**321.

3

Role of Lipids in Energy Metabolism

A.M.T. Beenakkers, D.J. Van der Horst, and W.J.A. Van Marrewijk

1. Introduction

Most reviews of lipid metabolism in insects have covered all classes of lipid compounds. Since lipids are generally defined as substances poorly soluble in water but soluble in organic solvents, the authors had to deal with compounds with divergent physiological functions, e.g., phospholipids and pheromones. The subject of this chapter limits the lipoidal substances to be discussed to those that provide a direct source of metabolic energy, though we recognize that other important lipid classes, for example those contributing to (sub)cellular components, are also essential for metabolism.

As insect development and insect flight are the most energy demanding processes, this chapter will focus on these two phenomena. Flight will receive most emphasis, partly because of our own interest, but particularly as in the last 10–15 years substantial progress has been made in aspects of lipid metabolism during flight.

As the reader will perceive, one of the difficulties encountered in surveying the relevant literature is that experimental results may vary considerably among the various insect species and even within a species, owing to differences in developmental stage, adult age, sex, or nutritional state of the insect; wherever possible and necessary disparities in experimental data owing to these differences will be discussed.

A.M.T. Beenakkers, D.J. Van der Horst, and W.J.A. Van Marrewijk • Laboratory of Chemical Animal Physiology, State University of Utrecht, 3508 TB Utrecht, The Netherlands.

2. Lipids as an Energy Source in Development

Biochemical and physiological aspects of insect development have been reviewed by several authors (Fast, 1964, 1970; Chen, 1971; Agrell and Lundquist, 1973; Downer and Matthews, 1976). Within the scope of our subject, this section will deal with the role of lipids in supplying energy for embryonic and postembryonic development, as well as during oogenesis and diapause.

2.1. Embryogenesis

Studies on the role of lipids during embryogenesis have been restricted mainly to changes in the lipid content of insect eggs. Interpretation and comparison of the many diverse reports is difficult, as in many cases lipid content has been expressed as percentage of the wet weight of numbers of eggs, with no regard to any changes in water content during the embryonic period. In other studies, lipid contents are given as percentages on a dry weight basis, or as actual quantities per individual egg.

There is a wide variation in the amount of egg lipid between different species. Lipid contents given by Gilbert (1967a) for the eggs of nine insect species representing four different orders varied from 1.5% of the wet weight (*Lymantria dispar*) up to 12% (*Bombyx mori, Melanoplus differentialis*). Lipids amount to 13.9% of the wet weight of eggs of the aphid *Myzus persicae* (Strong, 1964) and to 18.4% of those of the house cricket *Acheta domesticus* (Lipsitz and McFarlane, 1970).

In a number of insects, the nature of the lipids in the eggs and changes in the lipid patterns during embryonic development have been studied. The lipid pattern in developing embryos of *Locusta migratoria* has been traced by Allais *et al.* (1964). In the newly laid eggs, 26% of egg dry weight is lipid. The acylglycerols, which account for 78.5% of the total lipids, show a decrease of 31.2% of the initial amount throughout the embryonic period. This fall is due entirely to the catabolism of triacylglycerol (TG) (which constitutes initially 83% of total acylglycerols), as monoacylglycerol (MG) and diacylglycerol (DG) remain unchanged.

Neutral lipids constitute the major lipid component in the eggs of many other insects, including those of *Periplaneta americana* (Kinsella and Smyth, 1966), *Leucophaea maderae* (Gilbert, 1967b), *Plodia interpunctella* (Yurkiewicz and Oelsner, 1969), and *A. domesticus* (Lipsitz and McFarlane, 1971). In these species 86% to 94% of lipid in the eggs is neutral lipid, of which TG is the predominant component (85% to 96%). The marked decrease in the level of neutral lipid during embryogenesis is attributable almost exclusively to the utilization of TG.

These and many other data indicate that neutral lipid, particularly

TG, is an important energy source for embryonic development in insects. This does not hold true, however, for the entire embryonic period of all insect species. In the Japanese beetle, *Popillia japonica*, glycogen was considered to be the main source of energy during the early part of embryogenesis (Rothstein, 1952). In the milkweed bug, *Oncopeltus fasciatus*, by far the greatest portion of the lipid decrease occurs during the last half of the embryonic period, and Babcock and Rutschky (1961) have calculated that lipids represent only one third to one half of the material oxidized for energy in eggs of this insect. The importance of glycogen for embryonic development of the mealworm, *Tenebrio molitor*, has been shown by Ludwig and Ramazzotto (1965). The decrease of glycogen was six times that of free lipid through the embryonic period of the mealworm.

In addition to the described chemical changes, further evidence for the dominating role of lipid in insect embryogenesis has been provided by some data on respiration. It has been shown in eggs of the grasshopper, *M. differentialis*, that a respiratory quotient (RQ) of 1.0 is maintained only during the first day (Boell, 1935), indicating that the embryonic period in which carbohydrate is the main fuel is very short. Throughout most of the embryonic period the RQ averages 0.71, which points to lipid as the main energy source. It has been calculated that in the egg of the grasshopper 75% of the oxygen upake is consumed in the oxidation of fat.

Babcock and Rutschky (1961) have discussed some fundamental differences between aquatic, acleiodic eggs and terrestrial, cleiodic eggs. In aquatic eggs protein would provide a large percentage of energy for development, nitrogenous wastes being excreted as urea or ammonia. In terrestrial eggs catabolism of lipid predominates, with a suppression of protein breakdown. Lipid oxidation yields more energy and more metabolic water than does protein catabolism and, in addition, has a sparing effect on protein, thereby preventing an accumulation of toxic nitrogenous wastes. The importance of the high potential energy of lipid and the high yield of metabolic water in lipid oxidation for *L. maderae* has been emphasized by Gilbert (1967b). Adult females of this species of cockroach carry the embryos within their brood sac for more than two months, and during this period there is a continuous threat of desiccation.

2.2. Postembryonic Development

Most holometabolous insects accumulate large amounts of lipid during larval development as energy reserves for later nonfeeding stages (cf. reviews of Fast, 1964, 1970). In general, the lipid content steadily increases until the middle of the last instar, followed by a sharp rise during the last part of this instar, resulting in young pupae having two or three times as much lipid as the early last instar. Storage of lipid also occurs in developing hemimetabolous insects. In *L. migratoria,* for instance, there is a small increase in fat body lipid in fourth instar larvae, followed by a much sharper increase during the fifth instar (Hill and Goldsworthy, 1968). It has been suggested that the buildup of fat body reserves might be influenced by the activity of the corpora allata. During the fourth instar, when juvenile hormone is present, glycogen reserves are accumulated, whereas in fifth instar larvae, in the absence of juvenile hormone, an apparent switch in metabolism favors the accumulation of lipid. As will be discussed later (see Section 3.7.1), juvenile hormone has an inhibitory effect on lipid synthesis.

Utilization of lipid during molting was shown in *B. mori*, the lipid content decreasing 15% to 20% during every molting period (Niemierko *et al.*, 1956). Lipid utilization during metamorphosis was observed in several insects, including *Malacosoma americana* (Rudolfs, 1926), *Musca vicina* (Levinson and Silverman, 1954), *M. domestica* (Pearincott, 1960), *Agria affinis* (Barlow, 1965), *Heliothis zea* (Lambremont and Graves, 1969), and *H. virescens* (Wood *et al.*, 1969). In contrast, Ludwig *et al.* (1964) reported a quantitative increase in lipid content in late metamorphosis of *M. domestica*. In *Lucilia cuprina* lipid appears to be the predominant energy source during early and late metamorphosis, whereas carbohydrate is used during midpupal life (D'Costa and Birt, 1966). Mitochondria from abdomens and thoraces of the developing adult *Lucilia* oxidize fatty acids, with maximal rates evident at adult emergence (D'Costa and Birt, 1969). Interestingly, flight muscles of mature adult flies catabolize only carbohydrates.

Changes in lipid content during development have been studied extensively in the silkmoth, *Hyalophora cecropia* (Gilbert and Schneiderman, 1961a). After a small drop from the first to the second instar, the total lipid content of this insect steadily increases throughout the larval life, so that at pupation lipids constitute between 5% and 7% of the insect wet weight. Male pupae contain 50% more lipid than female per gram wet weight. This sexual dimorphism in lipid content increases markedly during the pupal–adult transformation. Changes in total lipid and RQ patterns have shown that female pupae, in contrast to male, begin to catabolize lipid early in adult development, resulting in the male adult having three times the lipid content of the female (Domroese and Gilbert, 1964). In the adult moth, both sexes utilize lipid as the favored substrate. A similar sexual dimorphism in lipid utilization during metamorphosis has been shown in the silkworms *B. mori* (Niemierko *et al.*, 1956) and *Antheraea pernyi* (Demyanovski and Zubova, 1957). The higher lipid contents of adult males were considered to be peculiar to Lepidoptera (Gilbert and Schneiderman, 1961b; Gilbert, 1967a). As discussed later (see Section

3.2), this sexual dimorphism may be correlated with mating behavior, lipid as a fuel for flight being required particularly for the male insect. In contrast, it is assumed that in most feeding adults, the female contains more lipid than the male, as lipid is an efficient substrate for egg development. However, the phenomenon of higher lipid contents in males than in females may be more widespread than suggested above. Fast (1970) has mentioned that males in orders that utilize only carbohydrate for flight, e.g., Diptera, also have more fat than females. Moreover, in some Lepidoptera that are strong fliers, e.g., *Danaus plexippus* and *Loxostege sticticalis*, the female contains more lipid than the male.

Data on lipid contents of a great number of insect species have been listed by Fast (1964). Mean lipid content on a dry weight basis is higher in larvae than in adults. In general, neutral lipid (NL) forms the major part of the lipids in insects at all developmental stages (Fast, 1970), Contents of NL as high as 90% to 96% of total lipid have been reported in larvae of several insects. During metamorphosis and adult life this percentage remains fairly constant in some insects, e.g., the tobacco budworm, *H. virescens*, (Wood *et al.*, 1969) and the beetle *Lyctus planicollis* (Mauldin *et al.*, 1971), whereas in others, e.g., the fruit flies *Dacus oleae* and *Ceratitis capitata* (Madariaga *et al.*, 1970a,b), there is a decrease of the relative NL content. The NL formed only 66% of the total lipid in fifth instar larvae of *H. cecropia*, but rose to 91% in newly ecdysed pupae; in the pupal fat body the TG represent 91.5% of the NL (Stephen and Gilbert, 1970). TG has been found to be the predominant lipid class in the fat body as well as in the whole insect. Wimer and Lumb (1967) showed that the percentage of TG in the fat body of the blowfly, *Phormia regina*, is directly correlated with the state of development. In the early third instar larvae, TG represented only 39% of total lipid, but this percentage steadily increased up to about 80% at the prepupal stage.

Changes in the fatty acid composition of the NL during development have been investigated only in a few insects. Although some developmental changes occur, in most cases the significance of these changes is only partly understood. For instance, the palmitic/palmitoleic acid ratio (16:0/16:1) in the NL fraction of *C. capitata* shows a clear increase during the larva–pupa–adult transitions, whereas in the related species *D. oleae*, this ratio decreases (Madariaga *et al.*, 1970a,b). In *P. americana* the changes with age of the percentage of saturated fatty acids in the TG and free fatty acids of fat body and hemolymph suggest a preferential utilization of TG composed of saturated fatty acids by young adults (Nelson *et al.*, 1967).

Alterations in the fatty acid composition of *H. cecropia* during development, as well as the involvement of juvenile hormone in these changes, have been investigated by Stephen and Gilbert (1970). The unsaturated fatty acids increase from 77% of NL in the newly emerged first instar to 84% in the fifth instar and even 97% in the pupae and early pharate adults, and represent more than 90% of NL through adult life. In NL of the fat body the amount of palmitoleic acid doubles during the last three days of pharate adult life. At this stage, a conversion of polyunsaturated into saturated and monounsaturated fatty acids appears to occur. In addition, it was shown that when the titer of juvenile hormone is high, the concentration of saturated fatty acids is also high (e.g., in first instar larvae), whereas a low titer or absence of juvenile hormone coincides with low saturated fatty acid concentrations (e.g., in chilled pupae and early pharate adults). According to the hypothesis of Stephen and Gilbert (1970), polyunsaturated acids in the active compartment of the fat body (cf. Beenakkers and Gilbert, 1968) are degraded to acetyl CoA, from which saturated acids are resynthesized. Juvenile hormone may act at this point of resynthesis to block desaturation. The enhanced concentrations of saturated fatty acids would then result in the release of DG, which is preferentially composed of saturated fatty acids.

2.3. Oogenesis

Because lipids are the predominant energy source during embryonic development of insects, sufficient amounts of lipid must be stored in the developing oocytes during oogenesis. Thus, in the ovoviviparous cockroach *L. maderae* the lipid content of the ovary rapidly increases during oogenesis, with a maximum level attained prior to ovulation and fertilization. Gilbert (1967b) has shown that there is a decrease in fat body lipid during the later stages of oogenesis that precedes the peak of oocyte lipid content. This decrease in fat body lipid is prevented by castration. Since egg maturation appeared to stimulate feeding, and since isolated ovaries (with maturing oocytes) were capable of incorporating labeled palmitate into acylglycerol and phospholipid, Gilbert (1967b) suggested that the storage of lipid in the oocytes may be a result of three processes: transfer of substrate from the fat body, transport of digested food directly to the ovary, and synthesis of lipid by the ovary. According to Martin (1969), the ovaries of the bug *Pyrrhocoris apterus* must obtain lipid both from the fat body and from the diet, neither one of which alone provides a sufficient amount. When the dietary source of lipid is limited, e.g., when females are not allowed to feed during vitellogenesis, mobilization of lipid from the fat body increases.

The lipid content of hemolymph vitellogenin, the predominant yolk protein precursor, varies among insects between about 7% and 16% (Engelmann, 1979). Incorporation of vitellogenin by the oocytes will contribute to the lipid reserves of the egg. Chino *et al.* (1977a) have shown that the DG content of egg vitellogenin of *Philosamia cynthia* is considerably lower than that of hemolymph vitellogenin (also termed LP-II, see Section 3.3.3), whereas TG, the major yolk lipid , comprising more than 90% of total lipid, is present in egg vitellogenin but absent in hemolymph vitellogenin. They therefore assumed that the uptake of vitellogenin into oocytes is coupled with the esterification of DG to form TG, and that TG once formed leaves the vitellogenin molecule to contribute to the formation of oil droplets in yolk. However, in view of the high lipid content of mature oocytes of *H. cecropia* (28.7%), Chino *et al.* (1977b) concluded that the lipid complement of the egg is derived only in part from the lipid associated with vitellogenin. The major effector of lipid transfer from the fat body to the developing ovary was considered to be lipoprotein-I (LP-I) (see Section 3.3.3; and chapter by Chino in this volume). Egg LP–I contained only 3.6% of the DG content of hemolymph LP-I. In addition, LP–I isolated from the egg retained the physiological capacity of hemolymph LP–I to take up DG from fat body. Thus, it was suggested that LP-I is the major source of lipid for vitellogenesis, functioning as a true carrier protein serving to transport DG from the fat body to the ovary.

2.4. Diapause

Diapause is a delay in development which results in the synchronization of active stages of morphogenesis with suitable environmental conditions. This adaptive phenomenon enables the insect to live in areas that at times are unsuitable for it. In most diapausing insects only one stage, characteristic of the species, enters diapause, although the developmental stage may differ among species. Diapause is characterized by a considerable reduction in metabolism; diapausing larval and adult forms do not feed. The onset of diapause must therefore be preceded by a buildup of food reserves.

The energy required during periods of diapause may be largely derived from the oxidation of fatty acids. Lambremont *et al.* (1964) have reported for the boll weevil, *Anthonomus grandis*, that after a period of feeding nondiapausing adults contained 6% to 10% lipid (40% to 60% of which is TG), whereas in diapausing adults the lipid content was 18% to 25% (75% to 85% TG). In a study on lipid composition of the face fly, *Musca autumnalis*, the wild form overwintering by adult diapause, Pitts and Hopkins (1965) have shown that the total lipid concentration of wild diapausing flies increases about seven times over that of nondiapausing flies. TG is the major lipid storage form of diapausing face flies, its content being nine times higher in diapausing than in nondiapausing flies at 15 days after ecdysis (Valder *et al.*, 1969). There is also a large increase in free fatty acids between day 8 and day 15, corresponding to the periods of most rapid total lipid buildup and TG synthesis in diapausing flies. Accumulation of lipid by diapausing insects has been reported for several other species, as reviewed by Downer and Matthews (1976).

An insect exhibiting hormonal control of lipid accumulation for diapause periods is *B. mori*. When the so-called diapause hormone is present in the silkworm during the period of oogenesis the eggs will enter diapause. This diapause hormone is secreted by the neurosecretory cells of the subesophageal ganglion of silkworm pupae, and its release into the hemolymph is regulated by the brain via connectives to the subesophageal ganglion (Fukuda and Takeuchi, 1967). Diapause eggs are characterized by a higher lipid content than nondiapause eggs. In the presence of diapause hormone, the process of lipid deposition in the pupal ovary is accelerated or extended over a longer period (see the review by Steele, 1976). Removal of the subesophageal ganglion from pupae containing developing diapause eggs results in a decline in ovary lipid content and a corresponding increase in that of fat body. Implantation of subesophageal ganglia into pupae containing developing nondiapause eggs evokes an enhancement of lipid deposition in the ovary and a concomitant fall in fat body lipid. It was suggested that the diapause hormone may stimulate the uptake of hemolymph lipids by the ovary. In addition, activation of fatty acid synthesis by diapause hormone may also contribute to the increase in ovarian lipid. The effect of diapause hormone on lipid synthesis, however, appears to be secondary to its influence on carbohydrate metabolism (Steele, 1976).

3. Utilization of Lipids during Flight

3.1. Historical Perspectives

Studies on Hymenoptera and Diptera, some of them performed more than 40 years ago, demonstrated that energy for flight is exclusively derived from carbohydrates. During flight, *Apis mellifera* (Jongbloed and Wiersma, 1934) and *Drosophila* (Chadwick, 1947) display a RQ equal to unity. Flight muscle intensity in the honeybee is related to the carbohydrate concentration of the hemolymph (Beutler, 1936), and in flight-exhausted *Drosophila* glucose feeding restores the capacity for flight within 30–45 sec (Wigglesworth, 1949).

In 1940, Fulton and Romney indicated in an indirect way that lipid might also be used as an energy source for flight activity; in adult beet leafhoppers, *Eutettix tenella*, when present in their breeding area, fat content was about 40% of body dry weight, whereas specimens taken over 200 miles away, probably after migratory flight, contained only about

9% lipid. A similar phenomenon was observed by Beall (1948) in the monarch butterfly, *Danaus plexippus*, performing an autumnal migration from northern areas of the United States to southern regions.

Direct evidence for participation of lipid in insect flight was furnished by Krogh and Weis-Fogh (1951) and Weis-Fogh (1952) for the desert locust *Schistocerca gregaria*. During its first flight period of about 15 to 30 min, the RQ is about 0.82, but during prolonged flight this falls to 0.75. In the first half hour of flight mainly carbohydrate content decreases, whereas during the following hours lipid is the sole source of energy, as indicated by a gradual depletion of lipid reserves.

Zebe (1953, 1954) studied respiration of several Lepidoptera. During both rest and flight the RQ reached values of about 0.72, demonstrating that lipid is utilized in rest as well as flight. When resting *Agrotis pronuba* were fed or injected with glucose, the RQ increased with high values, sometimes even over 2.0, indicating synthesis of lipid from the administered carbohydrate. When these glucose-fed insects were forced to fly, the RQ immediately dropped to reach 0.7 again. These data indicate that glucose is not a direct substrate for flight muscles in the insects studied. Earlier, Kozhantshikov (1938) had shown that sugar feeding results in deposition of lipid in certain Lepidoptera, and he presumed that in adults the direct use of sugar for energy metabolism is not well developed.

However, it remained questionable how lipids participate in flight metabolism. Preparations of locust flight muscles failed to show oxidation of long-chain fatty acids. By contrast, acetate was respired rapidly (Rees, 1954), and therefore it was suggested that fatty acids are partly oxidized in the fat body, the resulting acetate being transported to and further metabolized in the flight muscles.

3.2. Substrate Utilization and Flight Duration

In a discussion on the relationship between the kind of substrate used during flight and the endurance of flight, Weis-Fogh (1967) indicated that animals that use lipid as a fuel can make long-range flights, whereas those that utilize carbohydrate are capable only of relatively short periods of continuous flight; this results from the higher energetic value of lipid compared to carbohydrate and the fact that glycogen, the main carbohydrate store in animals, contrary to fat, is always stored with an appreciable amount of water. It was calculated that 1 mg of lipid delivers as much energy as 8 mg of stored glycogen, emphasizing that for reasons of weight economy lipid is a more desirable substrate than carbohydrate. Moreover, lipid produces more metabolic water that does carbohydrate. Therefore, insects that remain airborne for many hours may be expected to utilize fat during continuous flight. This has been demonstrated for a number of insects (Beenakkers, 1969a; Cenedella, 1971; Kallapur and George, 1973; Brown and Chippendale, 1974; Van Handel, 1974). Prior to the period of migratory flight, the desert locust accumulates reserves of fat body lipid (Walker *et al.*, 1970). In *Aphis fabae* both fat and glycogen are consumed during tethered flight (Cockbain, 1961); glycogen is used during early flight, and fat is the principal fuel after the first hour. Flight capacity of this insect is directly related to the initial fat content. Fatigue occurs before all lipid reserves are depleted, small deposits remaining in the abdomen. This phenomenon was also observed by Van Handel (1974) in the moths *Spodoptera frugiperda* and *Heliothis zea*. *Homorocoryphus nitudulus vicinus*, occurring in large seasonal night-flying swarms, has large fat depots averaging over 400 mg/g dry weight, and Karuhize (1972) showed a relationship between flight duration and the decrease in total body fat. After 18–24 hr of flight, up to 200 mg lipid/g dry weight were consumed. *H. subvittatus*, a smaller nonswarming form of the same genus, and not an active flyer, has a much lower fat content (average about 150 mg/g dry weight), but in contrast to *H. n. vicinus*, a measurable amount of glycogen. It is supposed that only the latter substrate is used during the short flights of this species. Corresponding observations were made by Nwanze *et al.* (1976) in the adult cowpea weevil, *Callosobruchus maculatus*. Newly ecdysed nonflying adult males contain about 70 mg TG/g wet weight, whereas in the flying form, the TG content is 150 mg; it is suggested that the difference is related to flight ability. The adults do not feed, and thus must draw upon the energy reserves accumulated during larval development. This is true also for the silkmoth, *Hyalophora cecropia*. The newly ecdysed male adult has a high lipid content (Gilbert and Schneiderman, 1961b), which is believed to be correlated with mating behavior, as male moths fly long distances in search of a virgin female. The female undertakes only limited flights, and her lipid content is much less than that of the male. Carbohydrate content in *H. cecropia* is almost undetectable, and the RQ is low during adult life. The normal decrease in lipid content during the first week of adult life can be reduced by preventing flight activity (Domroese and Gilbert, 1964). Beenakkers (1969a) concluded that in lepidopterans that do not feed as adults, e.g., *Philosamia* and *Actias*, lipids are more advantageous than other fuel stores. Species that do feed in the adult stage, particularly those capable of prolonged flight (e.g., *Pieris* and *Agrotis*; Williams, 1958) will also rely heavily on lipid, but may use carbohydrate as well.

It is appropriate to indicate that most insects utilizing lipid for flight are capable of fatty acid oxidation in the flight muscles (see Section 3.6). As discussed by Bursell (Chapter 5) some insects that depend on proline oxidation for flight also rely on lipid stored in fat body (Bursell, 1977;

Weeda *et al.*, 1980). These "lipid metabolizing" insects will not be discussed further in the present chapter.

3.3. Lipid Mobilization and Transport

Insect species which are strongly dependent upon lipid as a fuel for flight metabolism [e.g., certain Lepidoptera and Orthoptera (Krogh and Weis-Fogh, 1951; Zebe, 1954)], mobilize lipid reserves in the fat body during flight and transport these by means of the hemolymph to the muscles.

Aspects of lipid mobilization and transport in insects have been recently covered by the reviews of Gilbert and Chino (1974), Bailey (1975), and Downer and Matthews (1976), and the topic will be considered in the chapter by Chino (Chapter 6). Therefore, the sections that follow will be restricted to the role of lipids in flight metabolism, and will focus on recent contributions to our knowledge rather than summarize the early findings.

3.3.1. Transport Forms of Fatty Acids during Flight

Data compiled by Bailey (1975) reveal that the major lipid store in insect fat body is TG, which always accounts for 83% or more of the neutral lipids. Of particular interest, therefore, is the predominance of DG as the major lipid component of hemolymph in all but a few species of insects examined thus far (Gilbert and Chino, 1974; Downer and Matthews, 1976; Downer, 1978). Tietz (1962, 1967) studied lipid release from the fat body of adult *Locusta migratoria*. When fat body was incubated *in vitro* with [^{14}C]palmitate, the label was rapidly incorporated into the acylglycerols of the fat body, and in a medium containing hemolymph, acylglycerol was released from the fat body. In experiments conducted with adults of *Melanoplus differentialis* and *Hyalophora cecropia*, Chino and Gilbert (1965) demonstrated the specific release of DG from fat body into hemolymph.

In the locust, the concentration of hemolymph DG is elevated severalfold during flight (Beenakkers, 1965, 1973; Tietz, 1967; Mayer and Candy, 1967, 1969a), whereas little change occurs in other lipid constituents of the hemolymph. Injection of an aqueous extract of corpora cardiaca, which contain adipokinetic activity (see Section 3.3.4), also induces elevation of the hemolymph DG level (Beenakkers, 1969b; Mayer and Candy, 1969b).

Thus, in representatives of two orders of insects (Lepidoptera and

Orthoptera) that use lipids for flight metabolism, DG may be important in lipid transport.

Indications for the actual use of DG as a substrate for flight muscle metabolism were obtained by the studies of Van Handel and Nayar (1972) on the moth *Spodoptera frugiperda*: During both rest and flight the turnover of the pools of DG appeared to be high enough to match the metabolic activities of the insects. However, as these data were inferred from homogenates of total insects, they do not allow conclusions to be drawn about the transport of lipids. In the locust *L. migratoria*, turnover of the DG pool in the hemolymph has been studied during flight and rest (Van der Horst *et al.*, 1978a). The elevated DG level stabilizes during prolonged flight, with the rate of DG mobilization matching the rate of utilization. Turnover rate during flight is accelerated more than eightfold over the resting value, and amounts to 3.4 mg DG/locust/hr. Based on the oxygen consumption of *Schistocerca gregaria* during flight (Weis-Fogh, 1952), Beenakkers (1965) calculated for *Locusta* that lipid is consumed at a rate of 4.1 mg/hr, if it is assumed that fatty acid is the sole respiratory fuel. Early in flight, when the oxidation of trehalose is more pronounced, lipid utilization is lower (Van der Horst *et al.*, 1978b), though the instantaneous labeling of respiratory carbon dioxide when flight activity is induced after pulse-labeling of the hemolymph DG indicates that DG contributes to the fuel for flight from the onset of flight (Van der Horst *et al.*, 1980). Although the above observations suggest that DG is specifically released from the fat body of all insects, the results of some authors indicate that free fatty acids or TG rather than DG are the principal forms of fatty acid transport in some insect species (see Bailey, 1975). Apart from the possibility that morphological anomalies and/or *in vitro* artifacts may account for some of these observations (Gilbert and Chino, 1974), either the developmental stage of the insects investigated (for example, larvae of the wax moth *Galleria mellonella*, Wlodawer *et al.*, 1966; Wlodawer and Lagwińska, 1967), or the use of nonlipid substrates for flight (for example, the cockroach *Periplaneta americana*, Cook and Eddington, 1967) render a discussion about these observations beyond the scope of this chapter.

So far as the nature of the hemolymph DG is concerned, it appears that 1,2-diacylglycerol is released from the fat body of the locust (Tietz *et al.*, 1975). Of particular interest is the recent finding that the 1,2-DG isolated from hemolymph of *Locusta* is stereospecific and has the *sn*-1,2-configuration with a remarkably high optical purity (Tietz and Weintraub, 1980; Lok and Van der Horst, 1980). Tietz and Weintraub (1980) analyzed stereospecifically the hemolymph 1,2-DG of locusts at rest, using phospholipase A_2 after phosphenylation of the DG, and obtained an optical purity of the *sn*-1,2 enantiomer of 80% to 90%. Lok and Van der Horst (1980) applied a ^{1}H-NMR method, using chiral shift reagents after conversion of the hemolymph 1,2-DG into 1,2-diacetyl-3-tritylglycerols, and found the optical purity of the *sn*-1,2-DG to be over 96%. The latter authors also investigated the configuration of the elevated hemolymph 1,2-DG after a two hour flight as well as after TG mobilization induced by injection of corpus cardiacum extract. Over 97% of the 1,2-DG appeared to be the *sn*-1,2 enantiomer in both these cases, indicating stereospecificity of the processes involved in their production.

3.3.2. Lipid Mobilization from Fat Body

Since lipid is stored in the fat body in the form of TG and transported to the flight muscles in the hemolymph in the form of DG, the fat body must contain lipolytic activity, involving hydrolysis of the long-chain fatty acylglycerol esters. Lipolysis is effected in the presence of lipases, a class of hydrolytic enzymes that has been defined in terms of specificity for long-chain fatty acylglycerol esters and capacity to hydrolyze the esters of emulsified glycerides at an oil–water interface (Brockerhoff and Jensen, 1974). Thus, lipases are distinguished from the more general esterases, which include all enzymes that catalyze the hydrolysis of ester linkages irrespective of chain length or solubility.

Gilbert *et al.* (1965) reported lipase activity in the fat body of *H. cecropia* with trioleylglycerol as a substrate. Chang and Friedman (1971) indicated the presence of lipases in the fat body of larval and pharate adult *Manduca sexta*; they did not, however, examine the adult stage.

Two lipases have been described in the fat body of *L. migratoria* (Tietz and Weintraub, 1978). An alkaline lipase preferentially hydrolyzes MG, whereas microsomes and the soluble supernatant of fat body homogenates contain an acid lipase, which is also active against DG and TG. Trioleylglycerol, which was used as substrate in the latter experiments, was, however, degraded but very slowly, with fatty acids formed as the major product. Compared to the rate of MG hydrolysis, the TG lipolytic activity is 10–15 times lower. Possibly the difficulties encountered in obtaining stable micellar TG dispersions may account for these *in vitro* results, which do not necessarily reflect the *in vivo* conditions. It is interesting to note that in the fat body of *P. americana*, which has little dependence upon lipid for flight metabolism, significant TG lipase activity was detected (Hoffman and Downer, 1979). DG was the primary end product, while lesser amounts of MG and free glycerol were formed. Though we confirm the results of Hoffman and Downer for *Periplaneta* fat body TG lipase, in *Locusta* fat body hardly any lipolytic capacity was revealed, the employment of their assay procedure notwithstanding (Van den Broek, Van Doorn, Van der Horst and Van Marrewijk, unpublished observations).

On the basis of observed differences in the fatty acid composition of fat body TG and hemolymph DG in *H. cecropia* and *L. migratoria*, Beenakkers and Gilbert (1968) and Beenakkers (1973) postulated the existence of two different glyceride pools in the fat body. According to this view, TG in the storage compartment is hydrolyzed mainly to MG, and the products are transported to the active compartment where DG with a specific fatty acid pattern is synthesized for the release into the hemolymph.

Now that the *sn*-1,2 configuration of the locust hemolymph 1,2-DG has been established (see Section 3.3.1), two alternatives for their biochemical formation by TG mobilization during flight become plausible. Tietz and Weintraub (1978, 1980) suggested that TG is degraded to 2-MG by a nonstereospecific lipase. The 2-MG would then act as substrate in the resynthesis of 1,2-DG. It was shown that fat body microsomal acyltransferase specifically utilized 2-MG (Tietz *et al.*, 1975); moreover, the microsomal enzyme appeared to be stereospecific and preferentially synthesized *sn*-1,2-DG (Tietz and Weintraub, 1980). However, other studies suggest that monoacyl cleavage of the stored TG is the primary route of 1,2-DG production in the fat body of *S. gregaria* (Spencer and Candy, 1976). The latter possibility, which may prevail also in the fat body of the cockroach (Hoffman and Downer, 1979), would require a TG lipase specific for the *sn*-3 position of glycerol. Lok and Van der Horst (1980) studied the positional distribution patterns of the fatty acids of *Locusta* fat body TG and hemolymph *sn*-1,2-DG. The occupation of the *sn*-2 position in both glycerides was in close agreement, whereas small differences existed between the *sn*-1 position of the DG and the primary positions of the TG. This would be evidence in favor of the 2-MG pathway. The evidence is not conclusive, however, since direct production of *sn*-1,2,-DG by a stereospecific TG lipase cannot be excluded if it is assumed that the fatty acid distribution in the positions *sn*-1 and *sn*-3 of the fat body TG is asymmetrical, as has been shown for the depot fat of the mealworm *Tenebrio molitor* (Brockerhoff *et al.*, 1966).

3.3.3. Lipoproteins and Lipid Transport

In vitro studies on the release of DG from insect fat body have shown that the presence of hemolymph in the incubation medium is a requisite for glyceride release (Tietz, 1962; Chino and Gilbert, 1964, 1965). Siakotos (1960) demonstrated in cockroach (*P. americana*) hemolymph that proteins carry neutral lipid and suggested that in the roach, lipid is transported by means of these lipoproteins. Since that time, a variety of lipoproteins have been identified in the hemolymph of several insect species, including three species of silkmoths and the migratory locust, and the first review on the biochemistry of insect plasma proteins has appeared (Wyatt and Pan, 1978). Purification and characterization of functionally recognizable hemolymph proteins, other than hemoglobin, was first achieved for two major lipoproteins in the hemolymph of the silkmoth, *Philosamia cynthia*, the DG-carrying lipoproteins I and II (DGLP-I and DGLP-II), by Chino *et al.* (1967, 1969). By analogy to mammalian blood lipoproteins, Thomas and Gilbert (1968, 1969) used the terms high density lipoprotein (HDL) and very high density lipoprotein (VHDL), respectively, for the equivalent lipoproteins from the related silkmoths *H. cecropia* and *H. gloveri*, which were isolated by preparative ultracentrifugation. To avoid confusion, DGLP-I (HDL) and DGLP-II (VHDL) have been simply renamed lipoprotein I (LP-I) and lipoprotein II (LP-II), respectively (Gilbert and Chino, 1974). It is LP-I that accepts DG from the fat body and functions to carry DG to other tissues, including the flight muscles and the ovary (Chino *et al.*, 1969, 1977b). LP-II does not accept DG from the fat body and has been identified as the female-specific vitellogenin (Chino *et al.*, 1976; see Section 2.3). Molecular weights as determined by sedimentation equilibrium are approximately 700,000 (LP-I) and 500,000 (LP-II) (Chino *et al.*, 1969; Pan and Wallace, 1974). Lipid constitutes 44% of the weight of LP-I and 10% of LP-II, and DG is the major acylglycerol present in both lipoproteins (Gilbert and Chino, 1974).

As far as we are aware, the dynamics of loading DG-carrying lipoproteins during muscular exercise, particularly flight activity, has been investigated only in locusts. In adult male *S. gregaria*, Mayer and Candy (1967) showed that hemolymph contains two electrophoretically separable lipoproteins (group A). After flight the lipid content of the group A lipoproteins is elevated, but, in addition, lipid appears in a second pair of proteins (group B) with lower electrophoretic mobility. The presence of lipid (chiefly DG) in the group B fraction suggested a relationship with the extensive mobilization of lipid from the fat body. Similar changes in hemolymph lipoproteins occurred after injection of extracts of corpora cardiaca (Mayer and Candy, 1969b). In resting *L. migratoria*, Mwangi and Goldsworthy (1977) reported that a yellow hemolymph lipoprotein (A) carries virtually all the DG. Injection of extracts of corpora cardiaca results in the appearance of a higher molecular weight fraction, A^+, to which the increased DG is bound, whereas fraction A contains less DG than in the resting condition. In addition, the absorbance of a nonlipid containing protein fraction (C) appears to have decreased. A possible combination between components of fractions A and C to form units of A^+ was therefore suggested; the resulting configuration may have a higher capacity for DG-loading than A. Work from our laboratory concerning the DG-transporting lipoproteins in adult male *Locusta* revealed that in addition to the lipoprotein fractions A and A^+, a third very high molecular fraction (0) is present; this binds DG at rest and after flight or injection of corpus cardiacum extracts (Van der Horst *et al.*, 1979). Some data are shown in Fig. 1. Lipoprotein A is composed of a DG-carrying yellow fraction and a blue protein fraction containing no lipid. At rest, the DG/protein ratio of the yellow A lipoprotein is 0.18, whereas this ratio of the O fraction is 0.78. After flight, the ratio of fraction A is slightly reduced, but the DG/protein ratio of A^+ is increased to 1.36 and that of the O lipoproteins to 1.10. Since total protein concentration in the hemolymph remains unaltered during flight, it may be supposed that the DG-carrying lipoproteins that are formed or elevated originate completely from the association of hemolymph proteins present in the resting locust. Obviously, the yellow A lipoproteins are implicated in the formation of the (yellow) A^+ lipoproteins; the role of other proteins remains unclear. Immunological studies on the relationship between the three lipoproteins, which are currently in progress in our laboratory in cooperation with De Loof (Leuven, Belgium), have demonstrated that anti-A serum also reacts with A^+ and O, indicating that A, A^+, and O share antigenic sites. The changes in hemolymph lipoproteins occurring after the onset of flight activity in the locust are of vital importance for both the transport capacity and the transport rate of the DG released from the fat body (Van der Horst *et al.*, 1979).

Peled and Tietz (1975) have isolated and purified a yellow lipoprotein of molecular weight 340,000 ± 50,000 from the hemolymph of (sex-unspecified) *Locusta*, and described an *in vitro* system for the release of DG from the fat body; the ability of the lipoprotein to stimulate DG release is only one third that of hemolymph. The addition of another (blue) hemolymph protein to the system, however, promoted the uptake of DG into the yellow fraction. Though a correlation of their findings with the changes *in vivo* is difficult, the yellow lipoprotein is probably identical with our lipoprotein A, for which we calculated a molecular weight of 450,000 (Van der Horst *et al.*, 1979). Gellissen and Emmerich (1980) report a molecular weight of 700,000–850,000 for the same fraction, though earlier a value of approximately 150,000 was estimated (Gellissen and Emmerich, 1977).

Analogy between DG transport in Lepidoptera and in *Locusta* has been postulated because lepidopteran LP-I exhibits properties similar to those of *Locusta* lipoprotein A, which has also been named LP-I (Gellissen and Emmerich, 1980); it is our opinion that the paucity of information available on these transport systems hardly merits such a generalization. In the few species of Lepidoptera examined so far, the transport of DG to the site of utilization during flight has still to be assessed.

In the locust, it is not yet clear why two lipoprotein fractions carry DG in the resting condition whereas at least three fractions are required

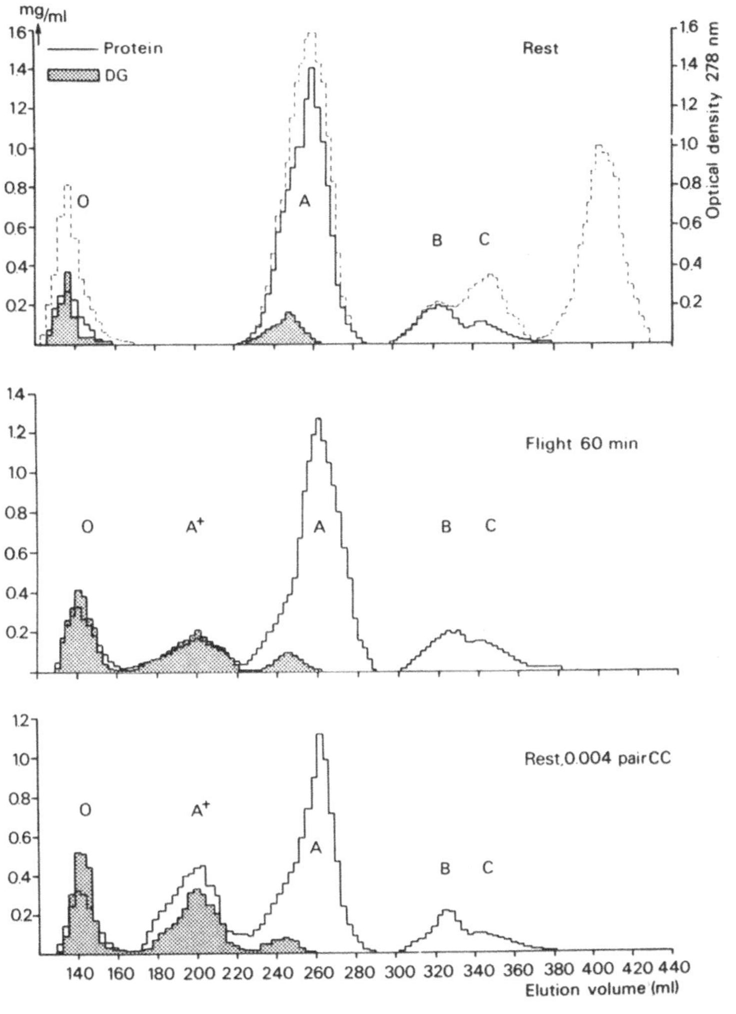

Fig. 1. Elution profiles of locust hemolymph (lipo)proteins. Hemolymph samples (1.0 ml) were collected from groups of adult male locusts at rest, after a 60 min flight, and 90 min after injection with an extract of the corpora cardiaca (CC) containing adipokinetic activity (0.004 pair corpora cardiaca equivalents/locust), and fractionated by gel filtration on Ultrogel AcA 22 (LKB Ltd.). Fractions were assayed for optical density (278 nm), protein, and DG content.

during flight. Studies from our laboratory indicate a rapid exchange of DG (labeled with [^{14}C]fatty acids) between the components of the system. Isolated and radioiodinated [^{125}I]lipoprotein A, when injected into the hemolymph of resting locusts, transfers a considerable amount of radiolabeled protein to the protein fraction of lipoprotein O (Van der Horst, De Keijzer, and Van Doorn, unpublished observations). The formation of higher molecular weight lipoproteins elicited by either flight activity or injection of extracts of corpora cardiaca may promote lipid transport and regulate catabolism, as has been described for mammalian plasma lipoproteins (Smith *et al.*, 1978).

3.3.4. Hormonal Influences on Lipid Mobilization and Transport

The regulation of lipid mobilization from insect fat body appears to be under neuroendocrine control. The existence of an adipokinetic hormone (AKH) located in the corpora cardiaca and concerned with flight activity was first suggested by the work of Beenakkers (1969b) and Mayer and Candy (1969b) on locusts. Aqueous extracts of corpora cardiaca elevate the DG level of the hemolymph and, simultaneously, changes occur in the hemolymph lipoproteins that are similar to those during flight (see earlier discussion). The site of action of AKH is the fat body (Mayer and Candy, 1969b), since release of DG from fat body *in vitro* is increased by the addition of corpus cardiacum extract. Extensive studies by Goldsworthy and colleagues have demonstrated that AKH is stored in and released from the glandular lobes of the locust corpus cardiacum (reviews: Goldsworthy *et al.*, 1972a; Goldsworthy and Mordue, 1974). Recently the hormone has been purified, characterized, and synthesized (Stone *et al.*, 1976; Broomfield and Hardy, 1977; Holwerda *et al.*, 1977), and appears to be a blocked decapeptide. In addition, a second adipokinetically active peptide has been isolated from *S. gregaria* corpora cardiaca (Carlsen *et al.*, 1979), the physiological role of which has not yet been assessed.

During flight of the locust, an increase in secretory activity of the intrinsic neurosecretory cells of the glandular lobe, namely, exocytosis of electron-dense neurosecretory granules, has been demonstrated (Rademakers and Beenakkers, 1977; Krogh and Normann, 1977) along with an elevation of the titer of the AKH in the hemolymph (Cheeseman *et al.*, 1976; Cheeseman and Goldsworthy, 1979). Stone and Mordue (1979) isolated the neurosecretory granules and demonstrated that they contain the AKH.

Results from our laboratory show that, in the locust, injection of corpus cardiacum extract raises the level of cAMP in the fat body (Gäde and Holwerda, 1976). Maximum increase occurs approximately five minutes after injection, and after ten minutes the cAMP level is returned to normal. Injection of the dibutyryl derivative of cAMP mimicked the lipid-mobilizing effect of corpus cardiacum extract. This may indicate that hormonal control of substrate mobilization is mediated by cAMP, which in turn may activate the enzymes involved in DG release (lipase) by way of protein kinase(s). Indeed, cyclic nucleotide-dependent protein kinase activity has been demonstrated in locust fat body (Beenakkers *et al.*, 1978; Van Marrewijk and Beenakkers, 1979). Chromatography of fat body extracts on hydroxyapatite results in the elution of a protein kinase which is activated by cAMP as well as by cGMP. Maximal activation with cAMP is higher and is reached at a lower nucleotide concentration than that obtained with cGMP; the apparent affinity constants for both nucleotides, however, are nearly identical. Despite these results *in vitro*, activation of fat body protein kinase *in vivo* after injection of corpus cardiacum extract could not be established, leaving the mechanism of mediation of the enhanced release of DG to be resolved. In *S. gregaria*, activation of fat body TG lipase also could not be detected after injection of the locusts with AKH (Spencer and Candy, 1976).

As locust AKH also shows hyperglycemic activity, though not in the locust, (Downer, 1972; Beenakkers *et al.*, 1978), it is interesting to note that AKH-stimulated activation of glycogen phosphorylase in locust fat body has been demonstrated (Van Marrewijk *et al.*, 1980a). Apart from a function in regulating fat body lipolytic processes, locust-AKH may have a direct effect on the process of DG-transporting lipoprotein formation. When the release of AKH from the corpora cardiaca during flight is prevented by experimentally elevating hemolymph concentrations of disaccharides (trehalose, sucrose) (Cheeseman *et al.*, 1976), neither the characteristic elevation of hemolymph DG concentration nor the formation of the flight-specific DG-carrying hemolymph lipoproteins is observed (Van der Horst *et al.*, 1979). Surprisingly, the turnover rate of the hemolymph DG is elevated considerably, indicating increased mobilization of DG from the fat body: Though the transport capacity of the DG is restricted to the lipoproteins A and O, the transport rate of the DG was about half the normal flight value thus greatly exceeding the resting value. Injection of locusts, flying under trehalose or sucrose loads, with extract of corpora cardiaca results in a rapid increase of both the content and the turnover of the hemolymph DG; this is accompanied by formation of the flight-specific hemolymph lipoprotein system capable of accepting the elevated lipid. Therefore, it was suggested that the primary effect of the increasing AKH titer at the onset of flight activity in the locust is to elicit the changes in association of the hemolymph lipoproteins that allow both the enhanced concentration and the increased transport rate of DG required for optimum energy supply for the flight muscles (Van der Horst *et al.*, 1979).

3.4. Flight Muscle Characteristics Pertinent to Fatty Acid Oxidation

The highly specialized function of insect flight muscles to perform mechanical work is clearly reflected in their morphological and biochemical composition and physiological qualities, which are far more specialized than those of other invertebrate and vertebrate muscles. The mitochondria are generally found in rows parallel to and between the myofibrils; mitochondrial volume occupies some 30% to 40% of the fiber volume (see for instance Ashhurst (1967) for *Lethocerus* species; Levenbook and Williams (1956) for *Phormia*, and Vogell *et al.* (1959) for *Locusta*) and equals the volume of the myofilaments. Flight muscles are, contrary to other insect muscles, rich in intracellular tracheoles, which are closely associated with the huge mitochondria (Edwards and Ruska, 1955). The metabolic rate of the working flight muscle is very high compared to that of other striated muscles (Weis-Fogh, 1952).

Fatty acid oxidation is a strictly aerobic process. The aerobic qualities of flight muscles anticipate the possibility of participation of lipids in energy metabolism. General mitochondrial metabolism and its control in insect flight muscles have been amply discussed by Sacktor (1975) and Crabtree and Newsholme (1975). The present section will focus mainly on the development of lipid oxidative capacities, particularly as expressed by development of fatty acid oxidative enzymes.

During organ development a progressive change and differentiation is observed. Since, in insects, developmental stages are sharply synchronized, it is possible to relate morphological and physiological structures to function. This holds true particularly for flight muscles, as only after the imaginal molt, when wing development is completed, the need for functional flight muscles becomes obvious. One of the structurally defined processes occurring during flight muscle development is the invagination of tracheoblasts into the muscle fibers. After invagination, the aerobic character of the muscle becomes prominent. (For a detailed description of tracheal development in the migratory locust, see Van den Hondel-Franken and Flight, 1980.) Electron microscopic and enzymatic studies together with investigations on protein biosynthesis demonstrated that development of the intrafibrillar tracheoles is followed by extensive mitochondriogenesis (Brosemer *et al.*, 1963; Kleinow *et al.*, 1970; Holmes and Keeley, 1975; Van Marrewijk *et al.*, 1980b; Keeley, this volume, Chapter 8). It is in this period that the final differentiation of the flight muscles towards the adult metabolic pattern takes place, as was first demonstrated by Brosemer *et al.* (1963) for *Locusta migratoria*. The moment of tracheolization, which coincides with highest lactate dehydrogenase activity throughout development, is followed by a period of differentiation in which the activity of mitochondrial enzymes belonging to the citric acid cycle and the glycerol-3-phosphate cycle is strongly increased; concomitantly the increased enzyme activity occurs with and is related to the rise in total mitochondrial membranes. Beenakkers *et al.* (1975) extended these observations on *Locusta* and also studied *Philosamia*. In the latter insect, as in the locust, the moment of maximum lactate dehydrogenase activity seems to be related to tracheolization, as suggested by structural studies on a related lepidopteran species by Bienz-Isler (1968). Concomitant with the reduction in lactate dehydrogenase activity are changes in the pattern of catabolic enzymes, which indicate that final differentiation occurs. In *Philosamia* and *Locusta* the activity of 3-hydroxy(= β-hydroxy) acyl CoA dehydrogenase (HOAD) (and in *Locusta* that of 3-ketoacylthiolase as well; Beenakkers, 1963) increases considerably. As indicated in Fig. 2, the rise in β-oxidative capacity of locust flight muscles during the period of larval muscle growth keeps pace with the increase in citrate synthase (CS), and thus with total mitochondrial membranes; however, during the differentiation phase, the increase in activity measured for HOAD is more accelerated and exceeds that of citrate synthase. The ratio HOAD/CS, which is about 0.6 in larval muscles, is raised to 1.2 in fully developed flight muscles. This emphasizes the progressive importance of lipid utilization in these muscles. In contrast, in the jumping muscles (extensor tibiae) of the locust the transition from larval to adult stage is not accompanied by marked changes in enzyme pattern; a small rise in anaerobic capacity coincides with the temporary suspension of oxygen supply via the interfibrillar tracheae. β-Oxidation remains subordinate to glycolysis.

3.5. Entrance of Lipids into the Flight Muscle

The mechanisms by which hemolymph lipids enter the flight muscle are virtually unknown. Therefore, this subject has not been covered by reviews dealing with lipid metabolism in insects.

Gilbert *et al.* (1965) have suggested that DG is separated from its carrier lipoprotein as it moves into the tissue. They have demonstrated that lipase in flight muscle of *Hyalophora cecropia* hydrolyzes DG at a rate that is five times that of TG hydrolysis, whereas in other tissues investigated, lipase activity is similar with both substrates. This may suggest that lipid enters the flight muscle as intact DG, after which it is hydrolyzed to fatty acids and glycerol. The possibility of extracellular hydrolysis of neutral lipid prior to uptake by muscular tissue has been discussed by Downer (1978). As support for the proposed mechanism, he mentioned the increased levels of fatty acids in hemolymph of the locust during flight. In vertebrates, plasma lipoproteins, altered by the enzymic action of lecithin cholesterol acyl transferase (LCAT) and li-

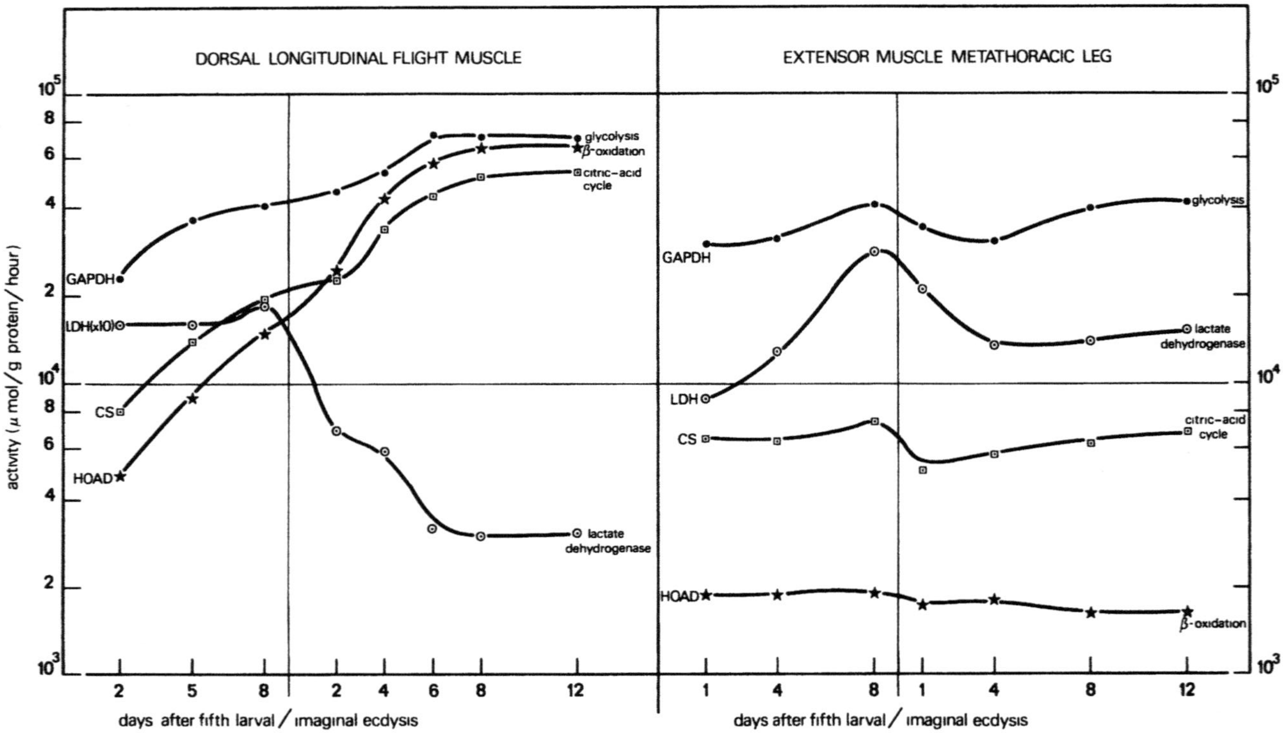

Fig. 2. Development of the main catabolic pathways in locust flight muscles and leg muscles. On the left side of each drawing the actual measured enzymes are indicated, on the right side the pathway they represent. GAPDH: glyceraldehyde phosphate dehydrogenase; LDH: lactate dehydrogenase; CS: citrate synthase; HOAD: 3-hydroxyacyl CoA dehydrogenase.

poprotein lipase and modified by processes of transfer and exchange, ultimately enter by endocytosis into a variety of cell types (see Smith *et al.*, 1978).

In light of the lack of data on the mechanisms of lipid uptake by flight muscle, it is not surprising that even less is known about the regulation of these mechanisms. Mayer and Candy (1969a) proposed that utilization of various fuels by flight muscle is largely controlled by the hemolymph concentrations of these fuels, and suggested alteration in the permeability of the muscle as another possible control mechanism for substrate utilization. From permeability studies on perfused working flight muscle preparations of *Locusta migratoria*, Robinson and Goldsworthy (1977) concluded that the use of DG in the muscle is stimulated by extracts of the glandular lobes of the corpora cardiaca. However, since this stimulating effect was prevented in the presence of 2-bromostearic acid (which blocks lipid oxidation), they suggested that the AKH has no direct effect on the rate of entry of DG into the muscle. The hormone may increase DG oxidation in the flight muscle primarily by stimulating the uptake of acyl groups into the flight muscle mitochondria, perhaps by acting on the inner carnitine acyl transferase. In contrast to the foregoing results, stimulation of DG oxidation in locust flight muscle could not be demonstrated by Candy (1978), despite employment of the same experimental procedure. From these few data it will be clear that further investigations are necessary to explain the process of lipid uptake by flight muscle as well as its regulation.

Fatty acids from hemolymph acylglycerols are made available for oxidation in flight muscle mitochondria by the hydrolytic action of lipases (see Section 3.3.2). In addition to hydrolysis by flight muscle lipases, extracellular hydrolysis by hemolymph lipases may be important as well.

Based on the inhibitory effect of high sodium chloride concentrations, Chang (1977) suggested that *Deilephila* hemolymph (as well as fat body and flight muscle) lipase is similar to mammalian lipoprotein lipase. A similar physiological function for cockroach hemolymph lipase (though this insect does not strongly depend on lipids during flight) and mammalian lipoprotein lipase has been proposed by Downer and Steele (1973). The two enzymes are structurally dissimilar, as, in contrast to the mammalian enzyme, cockroach lipase is inhibited by fluoride ions. The authors suggested that the source of hemolymph lipase is probably not the hemolymph itself, but the fat body or the gut. As the presence of high concentrations of DG in hemolymph is incompatible with the presence of continuously active hemolymph lipase, a mechanism of enzyme activation has been proposed that results in lipolytic activity only under particular physiological conditions (Hoffman and Downer, 1979). However, more research will be needed to establish the actual source of hemolymph lipase and to elucidate the regulatory mechanism of lipase activity.

Lipolytic activity in flight muscles has been reported for many insect species. Stevenson (1969) has demonstrated lipase activity with high specificity for MG in the flight muscle of four different moths. Although DG is the major lipid in the hemolymph of *Prodenia eridania*, Stevenson (1972) concluded from the very low DG lipase activity in the flight muscle that DG cannot be metabolized in this tissue. However, Stevenson's low values were obtained with DG emulsions, whereas DG lipase activity in moth flight muscle is much higher when its substrate is added as a part of its natural hemolymph lipoprotein complex (Gilbert and Chino, 1974). Crabtree and Newsholme (1972, 1975) have reviewed flight muscle lipase activities in 21 insect species representing eight different orders. High lipolytic activities were found with MG as a substrate, low with TG. In locust, waterbug, dragonfly, and some moths DG lipase activities are also high, which is consistent with the suggested role of DG as the main fuel for flight in these insects. Hoffman and Downer (1977, 1979) have shown that in the flight muscle (as in fat body and hemolymph) of *Periplaneta americana*, DG are the major end product of TG hydrolysis. By stereospecific analysis they were able to demonstrate that muscle and midgut homogenates produce a racemic mixture of the *sn*-1,2 and *sn*-2,3 enantiomers, whereas fat body and hemolymph accumulate preferentially the *sn*-1,2 enantiomer. As discussed in Section 3.3.1, the 1,2-DG in the hemolymph of *L. migratoria* also have the *sn*-1,2 configuration (Tietz and Weintraub, 1980; Lok and Van der Horst, 1980). These results suggest that degradation of DG in the flight muscle or hemolymph may be a stereospecific process, a particular enantiomer being hydrolyzed to fatty acids and glycerol by a specific DG lipase or lipoprotein lipase.

Glycerol produced in the flight muscle during lipolysis of acylglycerol could be converted to glycerol-3-phosphate by glycerol kinase. This enzyme is present in flight muscles of many insect species (Newsholme and Taylor, 1969; Crabtree and Newsholme, 1972). However, the activity of glycerol kinase in the flight muscle of locusts and moths does not seem sufficient to phosphorylate all of the glycerol produced by hydrolysis of DG. The concentration of glycerol in locust flight muscle increases threefold during a one hour flight (Worm and Beenakkers, 1980), and in the hemolymph, a tenfold increase was shown after the same flight period (Candy *et al.*, 1976). The latter authors have proposed that glycerol is released from the muscle into the hemolymph and transported to the fat body. In this tissue it provides an important source of glycerol-3-phosphate, which is required for the reesterification of fatty acids produced during conversion of TG to DG. The presence of highly active glycerol kinase in locust fat body (Zebe, 1959) supports the suggestion of the phosphorylation of glycerol.

If all of the glycerol produced in the flight muscle returns to the fat body, only part is required for reesterification of fatty acids. Candy *et al.* (1976) have shown that [^{14}C]glycerol injected into flown locusts is recovered from hemolymph not only as [^{14}C]-DG, but a substantial part is converted to [^{14}C]trehalose. Only two thirds of the trehalose used during a two hour flight can be supplied by the glycogen stores of the locust fat body (Van Marrewijk *et al.*, 1980a). The results of Candy *et al.* (1976) suggest that the glycerol not required for the reesterification process may contribute to the remaining trehalose consumed during locust flight.

3.6. Fatty Acid Oxidation in Flight Muscle

3.6.1. Pathways for Fatty Acid Oxidation

In the preceding sections, several aspects important to the processing of lipids for the eventual generation of energy have been discussed, leading to the concept that in a number of insect species the primary source of required metabolic energy for flight is lipid. This conclusion is based on RQ values, depletion of stored reserves of lipid during flight activity, phenomena of lipid mobilization and transport, and lipolytic activity of the flight muscles. Nonetheless, these important findings must not be taken to indicate that the utilization of fatty acids by insect flight muscle is understood completely. In the section that follows, processes involved in the breakdown of fatty acids are considered along with regulatory aspects.

The complete catabolism of fatty acids ultimately leads to the production of CO_2 and water. In general, lipid cannot be used as a fuel for muscular activity unless oxygen is available, i.e., there is no evidence for anaerobic utilization of lipids comparable to the conversion of carbohydrate into lactate (Crabtree and Newsholme, 1975). Thus, because of the nature of the substrate, there is an intimate association between the process of fatty acid degradation, the citric acid cycle, and the associated respiratory chain; the latter pathways being necessary for the first to be energetically efficient (Hochachka *et al.*, 1977).

As observed in mammalian tissues, the fatty acid moieties of lipids are broken down by the β-oxidative pathway into two-carbon units, which appear as acetyl CoA. The process of sequential shortening is outlined in Fig. 3. The initial step in fatty acid metabolism is the transformation of a parent fatty acid to the corresponding acyl CoA derivative, a reaction catalyzed by microsomal fatty acid thiokinases requiring ATP. Fatty acid

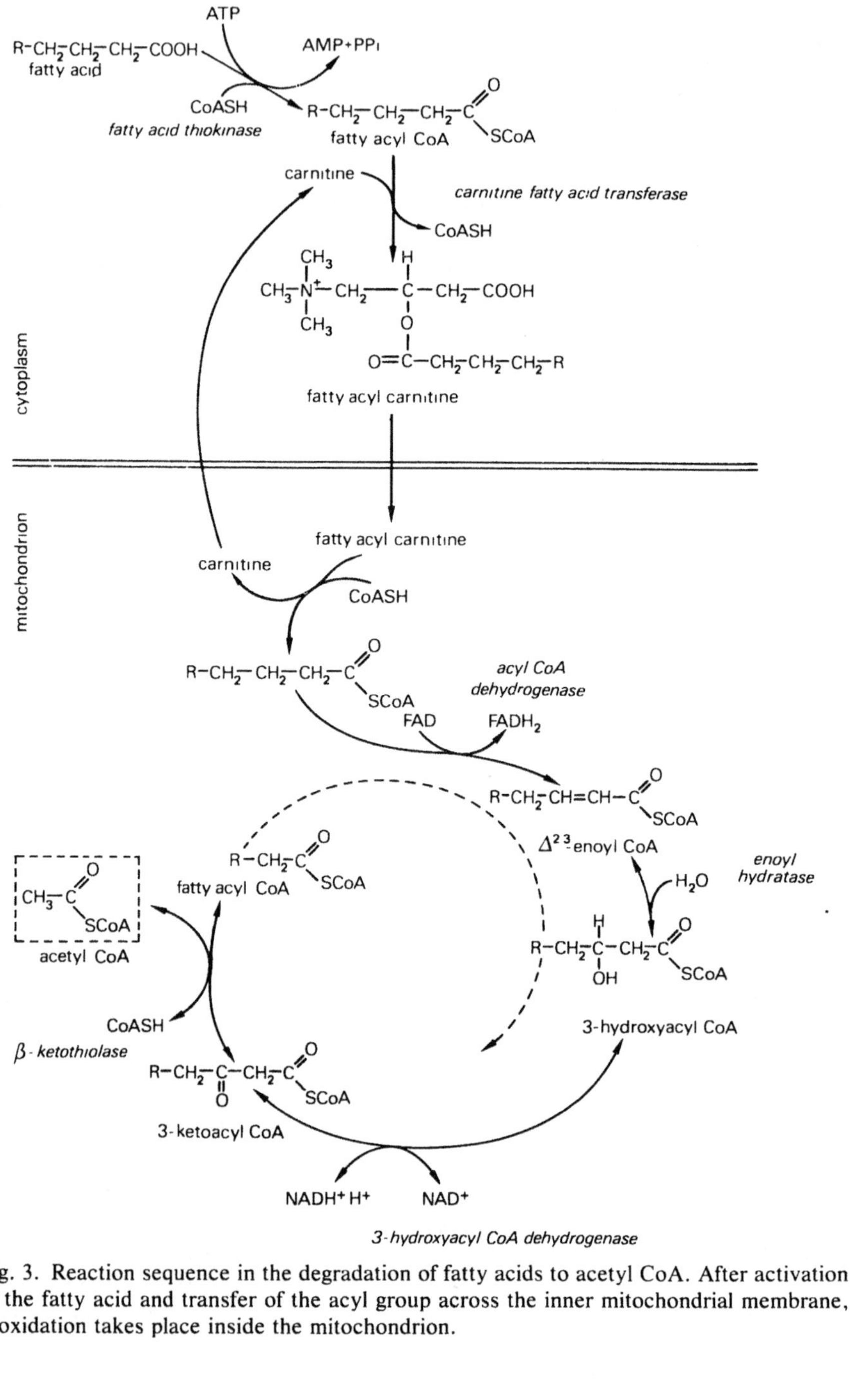

Fig. 3. Reaction sequence in the degradation of fatty acids to acetyl CoA. After activation of the fatty acid and transfer of the acyl group across the inner mitochondrial membrane, β-oxidation takes place inside the mitochondrion.

thiokinase activity is present in the flight muscle mitochondria of *Hyalophora cecropia* (Domroese and Gilbert, 1964) and *Prodenia eridania* (Stevenson, 1968). After transport of the fatty acyl moiety into the mitochondrion, the process of β-oxidation occurs. Of the four enzymes required for the β-oxidative pathway, 3-hydroxy-acyl CoA dehydrogenase (HOAD) and β-ketoacyl CoA thiolase (KAT) have been demonstrated in locust flight muscle mitochondria (Zebe, 1960; Beenakkers, 1963), whereas these enzymes are virtually absent in carbohydrate-utilizing insects (Beenakkers, 1969a). These results are strong evidence indicating the presence of the entire sequence of reactions in the flight muscles of insects that depend on lipids during flight. In view of the vigorous muscular activity associated with flight, the thoracic muscles of lipid-utilizing insects should yield mitochondria capable of rapidly oxidizing fatty acids; however, in early studies on respiration of flight muscle homogenates or mitochondrial preparations with long-chain fatty acids as substrate, disappointingly low rates of oxygen consumption were measured (Rees, 1954; Meyer *et al.*, 1960; Domroese and Gilbert, 1964). As indicated above, the initial activation of long-chain fatty acids is catalyzed by microsomal fatty acid thiokinases, whereas the enzymes of the β-oxidation reside within the inner membrane of the mitochondria. It has been recognized that in mammalian tissues this latter membrane is impermeable to fatty acyl CoA. Transport of fatty acyl CoA into the mitochondrion is effected by transfer of fatty acyl residues to carnitine, catalyzed by carnitine acyl transferase I. The acyl carnitine formed can cross the inner mitochondrial membrane in exchange for carnitine. On the matrix side of the inner mitochondrial membrane, the acyl carnitine is reconverted to carnitine and acyl CoA by carnitine acyl transferase II (Fritz, 1955; Fritz and Marquis, 1965; Brosnan *et al.*, 1973; Ramsay and Tubbs, 1975). In accord with this view, Beenakkers (1963, 1966) and Bode and Klingenberg (1964) have shown that oxidation of fatty acids in locust flight muscle homogenates was stimulated by the addition of carnitine. In isolated mitochondria, oxidation of long-chain fatty acids was observed only when they were administered as acyl carnitine esters, confirming that the prime activation of the acyl moieties normally occurs outside the mitochondria. It must, however, be stated that the high respiratory rates measured *in vivo* in free flying or tethered locusts have not been met in preparations *in vitro*. In *Locusta*, active carnitine palmitoyl transferase has been found (Beenakkers and Henderson, 1967; Beenakkers *et al.*, 1967; Crabtree and Newsholme, 1972). Of interest are the differences in carnitine acyl transferase activities between flight muscles of insects that rely on fatty acid oxidation during flight, such as locusts, and those of insects that utilize only carbohydrates. For instance, in *Apis mellifera* and *Phormia regina*, which belong to the latter group, carnitine palmitoyl transferase activities are negligible (Crabtree and Newsholme, 1972). As will be discussed later, the maximum catalytic activity of this particular enzyme *in vitro* has been related to the rate of lipid oxidation by muscles *in vivo* (Crabtree and Newsholme, 1972, 1975). However, these results conflict with the activities of carnitine acetyl transferase in carbohydrate-utilizing species. In the locust, the enzyme is very active, whereas it is absent from the flight muscles of the bee (Beenakkers and Klingenberg, 1964). But in blowfly flight muscles, which like those of the bee are deficient in fatty acid oxidation capacity, a very high level of carnitine acetyl transferase activity was found; it functions in carbohydrate utilization (Childress *et al.*, 1967) by acting in an acetyl CoA buffer system. If this transferase is directed mainly to the transfer of acetyl CoA units in lipid-oxidizing muscles, its function here could be to prevent large changes in the acetyl CoA-to-CoA ratio by providing an "acetyl sink," thus buffering the acetylation state of matrix CoA (Edwards *et al.*, 1974). Pyruvate dehydrogenase is inactivated by high concentrations of acetyl CoA (Pettit *et al.*, 1975). Therefore, in carbohydrate-utilizing muscles a temporary reduction of this intermediate by the transfer of the acetyl unit to carnitine is of importance (Childress *et al.*, 1967). In *Locusta,* the concentration of acetyl CoA in the flight muscle decreases during flight, whereas the acetyl carnitine concentration is raised (Worm *et al.*, 1980). No influence of the acetyl CoA-to-CoA ratio on β-oxidative capacity, however, has yet been demonstrated (see also Section 3.6.3).

Carnitine does not seem to be required universally among the insect species which are dependent on lipid for flight energy. Flight muscle mitochondria from the moths *P. eridania* and *Trichoplusia ni* can oxidize palmitate + malate very rapidly, with concomitant phosphorylation of ADP in the absence of carnitine; addition of carnitine had no effect on the rate of oxidation (Stevenson, 1966, 1968). Palmitoyl carnitine was not oxidized, and the presence of carnitine palmitoyl transferase was not demonstrated. Carnitine-independence is not universal within Lepidoptera; Hansford and Johnson (1976) reported that the very active oxidation of palmitate by flight muscle mitochondria of *Manduca sexta* is totally carnitine dependent. Perhaps, as suggested by Downer (1978), the apparent dichotomy of the requirement of carnitine in lipid-utilizing species may be related to the body temperature of the insect during flight, since the permeability of membranes for long-chain fatty acid increases with temperature.

3.6.2. Enzymic Patterns in Lipid-Oxidizing Flight Muscles

Generally, muscles that are capable of sustaining high metabolic rates always use lipid as the dominant metabolic fuel. As indicated before, there is an intimate association between the activities of the β-oxidation pathway and the citric acid cycle. In fact, since both pathways are located in the mitochondria, this integration leads to a close correlation between fatty acid oxidizing capacity and mitochondrial content (Hochachka *et al.*, 1977). Studies on the development of flight muscles of *Locusta migratoria* illustrate the integrity of the structural and metabolic processes involved (see Section 3.4). In the flight muscles of birds as well, a relationship exists between metabolic activity, capacity to utilize lipid for energy, and the concentration of various oxidative enzymes (George and Talesara, 1961). The systematic use of enzyme activities as an index of the activity of metabolic pathways has proved to be very suitable for comparative investigations (Vogell *et al.*, 1959; Klingenberg and Bücher, 1960; Pette, 1965). In this approach, the maximum catalytic activity *in vitro* of a given enzyme of a particular pathway is indicative of the rate of operation of that pathway in the intact muscle. The maximum catalytic activities of a number of enzymes are in approximate proportion in many tissues, and therefore it is assumed that they are also proportional to the maximum rate of operation of the pathway (Pette, 1965, 1966; Staudte and Pette, 1966; Beenakkers *et al.*, 1967; Beenakkers, 1969a). For example, in insect flight muscles HOAD has been used as an index enzyme for β-oxidation, glyceraldehyde-3-phosphate dehydrogenase (GAPDH) and lactate dehydrogenase (LDH) for general and anaerobic glycolytic pathways respectively, citrate synthase (CS) for the citric acid cycle, and glycerol-3-phosphate dehydrogenase (GDH) for the glycerol-3-phosphate cycle (Beenakkers, 1969a). In species depending on oxidation of only carbohydrates during flight, (for example, *A. mellifera*), activities of GAPDH, GDH, and CS in the flight muscles are very high, whereas HOAD activity is very low. The metabolic pattern of exclusively lipid-utilizing lepidopterous species, such as *Philosamia cynthia*, show extremely high HOAD and CS activities, and a third type of flight muscle, in which energy generation may be obtained by oxidation of both lipid and carbohydrate (for example, *L. migratoria*), is characterized by high activities of HOAD, GAPDH, GDH, and CS (Beenakkers, 1969a). In all types of flight muscles LDH activity is extremely low, emphasizing the fundamental aerobic properties of the muscles. The use of HOAD as an index enzyme for fatty acid oxidation has been questioned by Crabtree and Newsholme (1972, 1975), as maximum activities of this enzyme are greater than those of lipase and carnitine acyl transferase. Indeed, it has been shown that relative to cytochrome a content, the quotient of oxygen consumption of isolated mitochondria and the activity of carnitine palmitoyl transferase equals unity for locust flight muscles as well as for various organs of the rat (Beenakkers *et al.*, 1967). However, reliance on the activity of carnitine palmitoyl transferase to estimate the rate of fatty acid utilization in muscle, as suggested by Crabtree and Newsholme (1975), may lead to erroneous conclusions, in view of the carnitine-independent translocation of acyl moieties observed in some species of moths.

3.6.3. Regulatory Aspects of Fatty Acid Oxidation

The flight muscles of insects are the most active muscles known, and, consequently, they display exceptionally high rates of fuel utilization; this is facilitated by both metabolic specializations (such as a high oxidative capacity) and structural features (such as an elaborate conduit of tracheoles supplying oxygen) (see Sacktor, 1965, 1970). From a comparative viewpoint, these muscles offer advantages for studying regulation of metabolic processes; however, the sequence of events required to facilitate fatty acid oxidation in the transition from the resting condition to that of high muscular activity are only partially understood. In insect flight muscles, lipids as well as carbohydrates are oxidized completely; thus, the tricarboxylic acid cycle may be regarded as an extension of the pathway from lipid to acetyl CoA and, similarly, as an extension of the pathway from carbohydrate to pyruvate (Crabtree and Newsholme, 1975). This suggests that the rate of the tricarboxylic acid cycle may be regulated by the rate of formation of acetyl CoA from either lipid or carbohydrate. In blowfly (*P. regina*) flight muscles, which utilize carbohydrate exclusively (Sacktor, 1975), pyruvate oxidation is the sole source of acetyl CoA. Control of pyruvate oxidation is exerted primarily at the level of the dehydrogenases (Hansford, 1975). In the resting state, a high mitochondrial ATP/ADP ratio inhibits the NAD-linked isocitrate dehydrogenase, which eventually results in an accumulation of acetyl CoA and an inhibition of pyruvate dehydrogenase. Upon activation of respiration, these effects are reversed (Johnson and Hansford, 1975). In species that are known to oxidize fatty acids during flight activity, the question arises as to whether metabolic control is also exerted primarily upon the tricarboxylate cycle or whether control is directly upon the β-oxidative pathway supplying acetyl CoA to the tricarboxylate cycle. Hansford and Johnson (1976) determined in flight muscle mitochondria of *M. sexta* the steady state concentrations of certain key metabolic intermediates involving CoASH and CoA thioesters and their response to changes in flux through the metabolic pathway. They demonstrated that the limitation of carnitine palmitoyl oxidation in the resting state is at the level of the tricarboxylate cycle, implicating—as in *Phormia*—isocitrate dehydrogenase as a primary control point.

In insects that rely on both carbohydrate and lipid oxidation for flight muscle energy generation, a successive utilization of these substrates has been shown. Interestingly, despite the continuing availability of trehalose, which is the main substrate early in flight, oxidation of fatty acids derived from DG supplies the major portion of the energy requirements of the flight muscles during prolonged flight. However, regulatory aspects of this change in substrate consumption, are still unknown. In mammalian heart muscle, carbohydrate consumption is inhibited in the presence of long-chain fatty acids (Newsholme *et al.*, 1962). The first irreversible step in glycolysis is catalyzed by phosphofructokinase, the activity of which is inhibited by high levels of both ATP and citrate, thus providing a site of reciprocal control between glycolysis and β-oxidation (Newsholme and Randle, 1964; Hochachka *et al.*, 1977). Phosphofructokinase isolated from *Schistocerca* flight muscles, however, is not inhibited by citrate (Walker and Bailey, 1969). Ford and Candy (1972), working with perfused locust thoracic muscle preparations, confirmed the latter observation, and proposed trehalase, hexokinase, phosphofructokinase, aldolase, and pyruvate kinase to be control enzymes in the muscular tissue. The relative importance of some of the control sites may be different, and all sites may not operate under all physiological conditions. Recent work from our laboratory determined the concentrations of metabolites involved in carbohydrate oxidation, along with the concentrations of adenine nucleotides, during a 2 hr flight of *Locusta* (Worm and Beenakkers, 1980). Although a number of metabolites showed initial or enduring changes, with respect to the supposed mutual relationship between reduction of trehalose utilization and elevation of DG consumption, no clear indications of possible regulatory points in the glycolytic pathway were obtained. Therefore, it was proposed that a possible increase in the concentration of acetyl CoA derived from the β-oxidative pathway might inactivate pyruvate dehydrogenase (see Section 3.6.1). Changes in the concentrations of CoASH, carnitine, and their acyl derivatives in *Locusta* flight muscles during a 2 hr flight period were investigated to assess a possible regulatory role of these intermediates; the concentration of acetyl CoA decreased sharply immediately after the onset of flight, whereas the CoASH level remained relatively constant (Worm *et al.*, 1980). Thus, in the reciprocal control of acetyl CoA production by glycolysis and β-oxidation, either regulation must be exerted by other steps in the pathways, or other factors must be involved. Long-chain fatty acyl carnitine concentration in locust flight muscle increases gradually during flight (Worm *et al.*, 1980), suggesting that the rate of acyl translocation across the inner mitochondrial membrane limits the rate of long-chain acyl carnitine oxidation; however, the activity of carnitine palmitoyl transferase was calculated to be sufficient for the fatty acid respiratory rate of the flight muscles *in vivo*. In this context it is worth noticing that Robinson and Goldsworthy (1977) proposed that AKH stimulates the entry of fatty acyl groups into locust flight muscle mitochondria; perhaps by the action of the hormone on carnitine acyl transferase II (see also Section 3.5). Since Candy (1978) was not able to confirm the stimulation of the DG oxidation in perfused locust flight muscles, further investigation is needed before a physiological role of AKH on the regulation of flight muscle metabolism can be accepted.

3.7. Endocrine Control of Lipid Utilization and Flight Performance

Although a number of (neuro)hormones influence metabolism in insects, two are particularly involved in lipid metabolism, viz. juvenile hormone and adipokinetic hormone. Juvenile hormone is considered to participate in a variety of processes related to lipid metabolism during developmental, flight and migratory, and reproductive periods. AKH is involved principally in substrate mobilization and transport during flight (see Section 3.3.4). Several recent reviews have discussed hormonal influences on lipid metabolism in insects (Gilbert and Chino, 1974; Bailey, 1975; Goldsworthy, 1976; Steele, 1976; Downer, 1978), often with particular emphasis on synthetic processes and biochemical aspects. We therefore want to focus on specific physiological implications of both hormones in relation to general flight performance.

3.7.1. Juvenile Hormone

The influences of juvenile hormone on lipid metabolism in relation to insect flight are mainly, although not exclusively, investigated in migratory insects. This is comprehensible since participation of lipids plays an important role in this group of insects. Johnson (1963, 1969) has stated that most migratory flight in insects is prereproductive, and thus is performed by sexually immature insects after the teneral period; though in some species (e.g., locusts) migration and oviposition may alternate. Particularly in female insects, a relationship between migratory flight activity and development of the ovaries is conceivable, as both physiological phenomena may compete for the same substrate. External factors such as temperature, food quality and quantity, day length, and crowding may affect the preoviposition period of insects (Johnson, 1963).

A lengthened preoviposition period can prolong migratory flight; however, a direct inverse relationship between flight capability and egg ripening owing to competition for stored substrate appears to be excluded. Ovariectomy of immature *Locusta migratoria* prevents the normal age-dependent peak in flight performance that occurs about 15 days after the imaginal ecdysis (Lee and Goldsworthy, 1976). The same operation in *Melanoplus differentialis* (Pfeiffer, 1945) and *Schistocerca gregaria* (Hill and Izatt, 1974) results in normal age-dependent depletion of accumulated reserves. This phenomenon is also observed after gonadectomy of male desert locusts (Odhiambo, 1966a).

Numerous investigations have revealed that allatectomy soon after adult emergence results in a considerable accumulation of lipid. This has been shown, among others, in *M. differentialis* (Pfeiffer, 1945), *Drosophila* (Vogt, 1949), *Calliphora erythrocephala* (Thomsen, 1952), *Periplaneta americana* (Bodenstein, 1953), *S. gregaria* (Odhiambo, 1966a; Hill and Izatt, 1974), and *L. migratoria* (Strong, 1968a,b; Beenakkers, 1969b), thus including carbohydrate as well as lipid oriented insects. Most studies on the influence of corpora allata on lipid metabolism in relation to flight have been performed in locust species. Odhiambo (1966a) reported that the rise in lipid stores due to allatectomy could be reversed by reimplantation of corpora allata. In normally developing adult locusts, lipid accumulates during the period of somatic growth and slowly decreases with age after this period. Allactectomy does not alter the rate of accumulation, but as demonstrated in both *L. migratoria* (Strong, 1968b) and *S. gregaria* (Hill and Izatt, 1974), the operation allows a continuation of the lipid accumulation process after the period of somatic growth. The accumulation may be the result of several factors, including increased, or constantly high, feeding activity, decreased lipid utilization, inhibition of DG release from the fat body, increased lipid synthesis in the fat body, or any combination of these factors.

Odhiambo (1966a) observed an identical feeding pattern in both sham-operated and allatectomized male desert locusts, and similar results were obtained by Hill and Izatt (1974) in female *S. gregaria*. Beenakkers and Van den Broek (1974) demonstrated with female *L. migratoria* that neither allatectomy nor implantation of corpora allata significantly changes the feeding pattern or the total amounts of food consumed and absorbed. Therefore, feeding activity can be excluded as a determining factor for the observed lipid accumulation.

Experimental results obtained thus far do not indicate that juvenile hormone has an effect on lipid release from fat body and transport via hemolymph. Neither allatectomy nor implantation of active corpora allata have a lasting effect on hemolymph lipid content (Beenakkers, 1969b; Hill and Izatt, 1974; Lee and Goldsworthy, 1975); however, turnover studies revealing the absolute amounts of lipid transported have not been performed. Strong (1968a) showed that locomotor activity of allatectomized locusts results in a decrease of fat body lipid. There are no indications that allatectomized locusts are unable to effect lipid release during flight, and injection of corpora cardiaca extract into operated locusts brings about the normal rapid increase in hemolymph lipid (Goldsworthy *et al.*, 1972b). Even 16 days after allatectomy, the influence of AKH can be demonstrated; this indicates that the absence of juvenile hormone, although having a profound influence on synthesis of vitellogenin, does not affect the formation of nonvitellogenic lipoproteins.

Strong emphasis has been placed on the relationship between juvenile hormone and locomotion. Odhiambo (1966b) showed that removal of the corpora allata from male desert locusts reduced spontaneous locomotor activity, and Wajc and Pener (1971) observed a significant, though not very pronounced reduction in flight activity in allatectomized male *Locusta*. Strong (1968b), however, reported no reduction in locomotor activity in *Locusta* deprived of corpora allata. Lee and Goldsworthy (1975) showed a decline in flight performance in *Locusta* males, allatectomized when immature (at least in the first week after the operation), but demonstrated that *Locusta* females increased flight activity after allatectomy (Lee and Goldsworthy, 1976) (Fig. 4). In normal females, a peak in flight performance is reached at about day 15 after imaginal ecdysis, but in operated locusts this peak occurs later in adult life. In allatectomized insects of both sexes, the decline in flight activity, normally noticed in the third week after adult molt, is delayed. If allatectomy is performed in mature locusts the period of optimal flight performance is prolonged. It has been hypothesized that juvenile hormone influences processes related to flight muscle development (Goldsworthy, 1976). Indeed, implantation of active corpora allata into young fifth instar larvae of *L. migratoria* reduces flight muscle development (Poels and Beenakkers, 1969), and fatty acid-oxidizing enzymes, together with other catabolic enzymes, are also reduced (Van den Hondel-Franken *et al.*, 1980). However, neither implantation of corpora allata late in the fifth instar nor allatectomy of young adults has any effect on the attainment of normal adult flight mus-

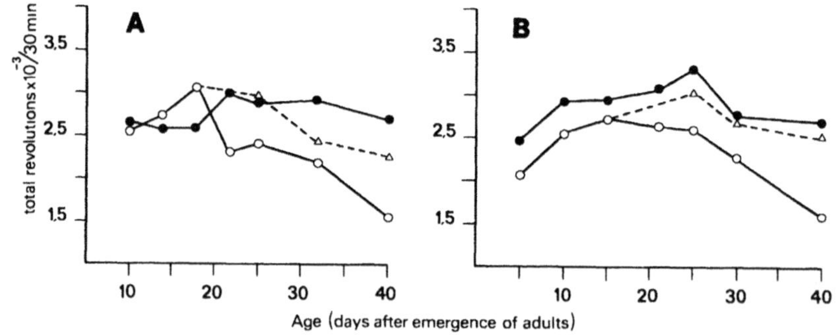

Fig. 4. Variations in flight intensity with age of control and operated male (A) and female (B) locusts: ○—○ operated control; ●—● allatectomized when immature; △– –△ allatectomized when mature. Locusts were flown, suspended on a roundabout (radius of 14 cm). Standard errors have been omitted for clarity (adapted from Lee and Goldsworthy, 1975, 1976).

cles (Beenakkers and Van den Broek, unpublished observations). Moreover, as already indicated, allatectomized locusts, when forced to fly, utilize fat body lipid (Strong, 1968a).

Wajc and Pener (1971) related the observed decrease in locomotor activity to the increased lipid content of allatectomized locusts. Thus, juvenile hormone might enhance flight performance, which in turn results in the utilization of lipid reserves. The various data cited above for related insects, and even for the same species, seem to negate this hypothesis. Moreover, as discussed by Rankin (1974), if juvenile hormone enhances flight activity and thus lipid consumption, it is difficult to understand that the same hormone also stimulates ovarian development, which is accompanied by a decrease in flight performance ("oogenesis–flight syndrome"; Johnson, 1969). The titer of the hormone is low during somatic growth and rises at the time of sexual development (Gilbert and King, 1973). Thus, it is possible that the concentration of the hormone is responsible for regulation of both specific physiological processes.

The most promising results regarding juvenile hormone and lipid metabolism have been obtained in experiments dealing with lipid biosynthesis in fat body. The present chapter is confined mostly to lipid metabolism as related to energy generation, i.e., catabolic processes, and the biosynthetic aspects will not be elaborated upon. Steele (1976) reviewed this aspect in relation to juvenile hormone. Suffice it to state that measurements of enzyme activities (e.g., Walker and Bailey, 1971) as well as incorporation experiments with radiolabeled fatty acids performed on whole insects (Vroman *et al.*, 1965, *P. americana*; Stephen and Gilbert, 1970, *Hyalophora cecropia*; Hill and Izatt, 1974, *S. gregaria*) and *in vitro* (Gilbert, 1967b) indicate that juvenile hormone inhibits lipid synthesis in fat body. These data concur with the observation that in the locust, allatectomy does not initiate the process of lipid accumulation in fat body, but rather prevents the termination of a previously initiated process. In the presence of an adequate titer of juvenile hormone, the fat body may switch from lipid to protein synthesis, particularly vitellogenic protein (Engelmann, 1970).

Thus, it is reasonable to suggest that juvenile hormone mainly interferes with synthetic processes of lipid metabolism. However, with regard to other lipid-related processes, e.g., flight, no definite statements can be made because of variation in experimental procedures. For example, in the implantation experiments the actual titer of juvenile hormone and the change in titer with time, are not known. Allatectomy is known to change the physiological age of the insects (Wajc and Pener, 1971), and this might influence other physiological processes. Finally, it is difficult to ascertain the consequences of intervention in one hormonal system for other (neuro)hormonal and neural systems. Therefore, it is pertinent to investigate the various parameters influencing lipid metabolism and flight during precisely defined developmental stages; moreover, in order to follow any direct or indirect influence of juvenile hormone on organ-specific metabolic properties, reliable *in vitro* systems will have to be established.

3.7.2. Adipokinetic Hormone

Results of various experiments indicate that hormone(s) of the corpora cardiaca influence flight performance of locusts. As the glandular lobe appears to be particularly involved, the influence is normally attributed to the adipokinetic hormone, which enhances lipid release from fat body. Weis-Fogh (1952) showed that flight speed, being high in the beginning of flight, is reduced during the next 10 min to about 70% of the speed reached in the first minute. These observations were corroborated and extended by Gewecke (1972, 1975), who, in locusts at about 30 days after the imaginal ecdysis, discovered a relationship between flight speed and wing beat frequency, the latter decreasing from 25 to 19 Hz during a 15 min flight period. Flight performance improves during adult development, and a constant relationship between wingstroke frequency and flight speed is demonstrated (Kutsch and Gewecke, 1979).

The change in flight pattern may be considered to reflect a qualitative change in substrate utilized by the flight muscles. Weis-Fogh (1952) showed that during the first 20–30 min of flight carbohydrate is the main substrate, whereas during prolonged flight lipid is preferred. This phenomenon has been substantiated and characterized by various authors (see Van der Horst *et al.*, 1980, and Section 3.3.1). It is generally assumed that during the period of high flight speed carbohydrate is oxidized, whereas lipid is utilized during the subsequent low speed flight.

AKH may be involved in regulating the delivery of substrate for flight and/or in influencing flight muscle metabolism. Starved locusts, which exhibit low levels of hemolymph trehalose and are unable to release AKH (Goldsworthy, 1969; Houben, 1976), fly poorly; injection of trehalose restores flight speed during the initial flight period (Goldsworthy and Coupland, 1974). The same authors also demonstrated that removal of the glandular lobes of the corpora cardiaca, although influencing flight speed minimally during the first 10 min of flight, strongly reduces flight performance in later flight. Injection of trehalose restores flight speed for about 30 min and, if corpora cardiaca extract is administered along with trehalose, flight performance is indistinguishable from that of controls. Locusts, flying with nonfunctional glandular lobes, due to severance of the nerves connecting the corpora cardiaca with the cerebral ganglion, exhibit a normal decrease in hemolymph carbohydrate, but DG concentration does not change significantly (Goldsworthy *et al.*, 1972b). Locusts injected with corpora cardiaca extract show reduced flight speed and reduced carbohydrate utilization; these observations led Robinson and Goldsworthy (1974, 1976) to suggest that the hormone not only influences DG mobilization, but also the utilization of DG and trehalose by the flight muscles, which in turn affects flight speed. Clegg and Evans (1961) demonstrated in the blowfly a correlation between hemolymph trehalose concentration and flight performance. Thus, the reduced carbohydrate utilization in the locust under hormone load may result in lower flight speed.

Injection of an emulsion of dipalmitoyl glycerol 30 min prior to flight results in reduced flight speed and carbohydrate utilization; if corpora cardiaca extract is also administered prior to flight, flight speed is restored and carbohydrate utilization is further reduced. These results, along with respiratory measurements on locust flight muscles *in vitro*, indicate that AKH acts directly on flight muscles by suppressing trehalose utilization (Robinson and Goldsworthy, 1974) and enhancing DG oxidation (cf. Robinson and Goldsworthy, 1977; see, however, Candy, 1978 and Section 3.5). However, the hormone may act indirectly on trehalose oxidation through its influence on lipid mobilization; the raised hemolymph DG content may compete with trehalose in flight muscle metabolism. This can explain why removal of the glandular lobe does not enhance flight speed in the first 10–20 min of flight, which would be expected if the hormone decreases trehalose oxidation. In normal flight carbohydrate provides most of the energy during the first 20–30 min of flight (Jutsum and Goldsworthy, 1976; Van der Horst *et al.*, 1978b, 1980). Moreover, injection of trehalose along with corpora cardiaca extract in cardiacectomized locusts restores normal flight, not only during later flight periods, when lipid oxidation is facilitated by the action of the hormone on fat body DG release and on fatty acid oxidation in the muscles, but also during the early stages of flight.

Normal flight pattern is also envisaged to be influenced by AKH inducing the utilization of lipid, which is accompanied by a decrease in flight speed. However, carbohydrate contributes an important part of the energy used in the flight period following the first minutes of high flight speed. If AKH reduces trehalose oxidation, as discussed above, a delay must occur after the initiation of flight before secretion of AKH can begin (Steele, 1976). However, Rademakers and Beenakkers (1977), showed that within the first 5 min of flight the hormone-producing glandular cells exhibit high secretory activity. At the same time, in the fat body a significant hormone-dependent rise in cAMP is measured (Gäde and Holwerda, 1976). However, about 15 min elapse before DG concentration in the hemolymph is significantly raised relative to the resting level (Van der Horst *et al.*, 1980), which supports the view that the relative concentrations of trehalose and DG determine their usage as energy substrates. This proposal retains the possibility that the hormone influences fatty acid oxidation in the flight muscles (Robinson and Goldsworthy, 1977). Moreover, a strict relationship between flight speed and the substrate utilized for flight is doubtful, judging from recent experiments in which locusts were flown for 30 min, rested for 2 hr, and then flown again for 30 min (Goldsworthy *et al.*, 1979). During both flight periods, flight speed was identical, viz. initially high and decreasing in the following minutes to a lower cruising speed. At the commencement of the first flight, the hemolymph shows normal high carbohydrate and low lipid levels, but at the start of the second flight, lipid concentration is still high and carbohydrate level is only half that of resting locusts. During the second flight, less than half of the diminished carbohydrate content is utilized. This indicates that the high initial flight speed in the second flight is supported by lipid oxidation.

The foregoing discussion leads to the conclusion that the nature of substrate oxidized in the flight muscles is unlikely to influence flight speed. It is more likely that the basic flight motor pattern is centrally programmed, with modifications resulting from reflex control by sense organs (Kutsch and Gewecke, 1979). Any causal effect of hormones on flight performance awaits further elucidation using pure hormones.

References

Agrell, I.P.S., and Lundquist, A.M., 1973, Physiological and biochemical changes during insect development, in: *The Physiology of Insecta*, Volume I (M. Rockstein, ed.), pp. 159–247, Academic Press, New York.

Allais, J.P., Bergerard, J., Etienne, J., and Polonovski, J., 1964, Nature et évolution des lipides au cours de l'embryogénèse de *Locusta migratoria migratorioides* L., *J. Insect Physiol.* **10**:753.

Ashhurst, D.E., 1967, The fibrillar flight muscles of giant water-bugs: An electronmicroscope study, *J. Cell Sci.* **2**:435.

Babcock, K.L., and Rutschky, C.W., 1961, Lipids in insect eggs: A review with new evidence from the milkweed bug, *Oncopeltus fasciatus* (Hemiptera, Lygaeidae),*Ann. Entomol. Soc. Am.* **54**:156.

Bailey, E., 1975, Biochemistry of insect flight. 2. Fuel supply, in: *Insect Biochemistry and Function* (D.J. Candy and B.A. Kilby, eds), pp. 89–176, Chapman and Hall, London.

Barlow, J.S., 1965, Composition of the fats in pupae of *Agria affinis* (Fallen) (Diptera: Sarcophagidae), *Can. J. Zool.* **43**:291.

Beall, G., 1948, The fat content of a butterfly, *Danaus plexippus* Linn., as affected by migration, *Ecology* **29**:80.

Beenakkers, A.M.T., 1963, Enzyme der Fettsäure-Oxydation in den Flugmuskeln von *Locusta migratoria* während ihrer Entwicklung, *Biochem. Z.* **337**:436.

Beenakkers, A.M.T., 1965, Transport of fatty acids in *Locusta migratoria* during sustained flight, *J. Insect Physiol.* **11**:879.

Beenakkers, A.M.T., 1966, The influence of carnitine on fatty acid oxidation in insect muscles, *Arch. Néerl. Zool.* **16**:535.
Beenakkers, A.M.T., 1969a, Carbohydrate and fat as a fuel for insect flight. A comparative study, *J. Insect Physiol.* **15**:353.
Beenakkers, A.M.T., 1969b, The influence of corpus allatum and corpus cardiacum on lipid metabolism in *Locusta migratoria, Gen. Comp. Endocrinol.* **13**:492.
Beenakkers, A.M.T., 1973, Influence of flight on lipid metabolism in *Locusta migratoria, Insect Biochem.* **3**:303.
Beenakkers, A.M.T., and Gilbert, L.I., 1968, The fatty acid composition of fat body and haemolymph lipids in *Hyalophora cecropia* and its relation to lipid release, *J. Insect Physiol.* **14**:481.
Beenakkers, A.M.T., and Henderson, P.T., 1967, The localization and function of carnitine acetyltransferase in the flight muscles of the locust, *Eur. J. Biochem.* **1**:187.
Beenakkers, A.M.T., and Klingenberg, M., 1964, Carnitine-coenzyme A transacetylase in mitochondria from various organs, *Biochim. Biophys. Acta* **84**:205.
Beenakkers, A.M.T., and Van den Broek, A.T.M., 1974, Influence of juvenile hormone on growth and digestion in fifth instar larvae and adults of *Locusta migratoria, J. Insect Physiol.* **20**:1131.
Beenakkers, A.M.T., Dewaide, J.H., Henderson, P.T., and Lutgerhorst, A., 1967, Fatty acid oxidation and some participating enzymes in animal organs, *Comp. Biochem. Physiol.* **22**:675.
Beenakkers, A.M.T., Van den Broek, A.T.M., and De Ronde, T.J.A., 1975, Development of catabolic pathways in insect flight muscles. A comparative study, *J. Insect Physiol.* **21**:849.
Beenakkers, A.M.T., Van der Horst, D.J., and Van Marrewijk, W.J.A., 1978, Regulation of release and metabolic function of the adipokinetic hormone in insects, in: *Comparative Endocrinology* (P.J. Gaillard and H.H. Boer eds.), pp. 445–448, Elsevier/North-Holland Biomedical Press, Amsterdam.
Beutler, R., 1936, Über den Blutzucker der Bienen, *Z. Vgl. Physiol.* **24**:71.
Bienz-Isler, G., 1968, Elektronenmikroskopische Untersuchungen über die Entwicklung der dorsolongitudinalen Flugmuskeln von *Antheraea pernyi* Guer. (Lepidoptera), *Acta Anat.* **70**:524.
Bode, C., and Klingenberg, M., 1964, Carnitine and fatty acid oxidation in mitochondria of various organs, *Biochim. Biophys. Acta* **84**:93.
Bodenstein, D., 1953, Studies on the humoral mechanism in growth and metamorphosis of the cockroach, *Periplaneta americana, J. Exp. Zool.* **124**:105.
Boell, E.J., 1935, Respiratory quotients during embryonic development (Orthoptera), *J. Cell. Comp. Physiol.* **6**:369.
Brockerhoff, H., and Jensen, R.G., 1974, *Lipolytic Enzymes*, Academic Press, New York.
Brockerhoff, H., Hoyle, R.J., and Wolmark, N., 1966, Positional distribution of fatty acids in triglycerides of animal depots fats, *Biochim. Biophys. Acta* **116**:67.
Broomfield, C.E., and Hardy, P.M., 1977, The synthesis of locust adipokinetic hormone, *Tetrahedron Lett.* **25**:2201.
Brosemer, R.W., Vogell, W., and Bücher, T., 1963, Morphologische und enzymatische Muster bei der Entwicklung indirekter Flugmuskeln von *Locusta migratoria, Biochem. Z.* **338**:854.
Brosnan, J.T., Kopec, B., and Fritz, I.B., 1973, The localization of carnitine palmitoyltransferase on the inner membrane of bovine liver mitochondria, *J. Biol. Chem.* **248**:4075.
Brown, J.J., and Chippendale, G.M., 1974, Migration of the monarch butterfly, *Danaus plexippus*: Energy sources, *J. Insect Physiol.* **20**:1117.

Bursell, E., 1977, Synthesis of proline by fat body of the tsetse fly (*Glossina morsitans*): Metabolic pathways, *Insect Biochem.* **7**:427.

Candy, D.J., 1978, The regulation of locust flight muscle metabolism by octopamine and other compounds, *Insect Biochem.* **8**:117.

Candy, D.J., Hall, L.J., and Spencer, J.M., 1976, The metabolism of glycerol in the locust *Schistocerca gregaria* during flight, *J. Insect Physiol.* **22**:583.

Carlsen, J., Herman, W.S., Christensen, M., and Josefsson, L., 1979, Characterization of a second peptide with adipokinetic and red pigment-concentrating activity from the locust corpora cardiaca, *Insect Biochem.* **9**:497.

Cenedella, R.J., 1971, The lipids of the female monarch butterfly, *Danaus plexippus*, during fall migration, *Insect Biochem.* **1**:244.

Chadwick, L.E., 1947. The respiratory quotient of *Drosophila* in flight, *Biol. Bull., Woods Hole, Mass.* **93**:229.

Chang, F., 1977, The presence of lipoprotein lipase activity and its relationship to lipid transport in the oleander hawkmoth, *Deilephila nerii, Comp. Biochem. Physiol.* **57B**:209.

Chang, F., and Friedman, S., 1971, A developmental analysis of the uptake and release of lipids by the fat-body of the tobacco hornworm, *Manduca sexta, Insect Biochem.* **1**:63.

Cheeseman, P., and Goldsworthy, G.J., 1979, The release of adipokinetic hormone during flight and starvation in *Locusta, Gen. Comp. Endocrinol.* **37**:35.

Cheeseman, P., Jutsum, A.R., and Goldsworthy, G.J., 1976, Quantitative studies on the release of locust adipokinetic hormone, *Physiol. Entomol.* **1**:115.

Chen, P.S., 1971, *Biochemical Aspects of Insect Development*, S. Karger, Basel.

Childress, C.C., Sacktor, B., and Traynor, D.R., 1967, Function of carnitine in the fatty acid oxidase-deficient insect flight muscle, *J. Biol. Chem.* **242**:754.

Chino, H., and Gilbert, L.I., 1964, Diglyceride release from insect fat body, *Science* **143**:359.

Chino, H., and Gilbert, L.I., 1965, Lipid release and transport in insects, *Biochim. Biophys. Acta* **98**:94.

Chino, H., Sudo, A., and Harashima, K., 1967, Isolation of diglyceride-bound lipoprotein from insect hemolymph, *Biochim. Biophys. Acta* **144**:117.

Chino, H., Murakami, S., and Harashima, K., 1969, Diglyceride-carrying lipoproteins in insect hemolymph. Isolation, purification and properties, *Biochim. Biophys. Acta* **176**:1.

Chino, H., Yamagata, M., and Takahashi, K., 1976, Isolation and characterization of insect vitellogenin. Its identity with hemolymph lipoprotein II, *Biochim. Biophys. Acta* **441**:349.

Chino, H., Yamagata, M., and Sato, S., 1977a, Further characterization of lepidopteran vitellogenin from haemolymph and mature eggs, *Insect Biochem.* **7**:125.

Chino, H., Downer, R.G.H., and Takahashi, K., 1977b, The role of diacylglycerol-carrying lipoprotein I in lipid transport during insect vitellogenesis, *Biochim. Biophys. Acta* **487**:508.

Clegg, J.S., and Evans, D.R., 1961, The physiology of blood trehalose and its function during flight in the blowfly, *J. Exp. Biol.* **38**:771.

Cockbain, A.J., 1961, Fuel utilization and duration of tethered flight in *Aphis fabae* Scop., *J. Exp. Biol.* **38**:163.

Cook, B.J., and Eddington, L.C., 1967, The release of triglycerides and free fatty acids from the fat body of the cockroach, *Periplaneta americana, J. Insect Physiol.* **13**:1361.

Crabtree, B., and Newsholme, E.A., 1972, The activities of lipases and carnitine palmitoyltransferase in muscles from vertebrates and invertebrates, *Biochem. J.* **130**:697.

Crabtree, B., and Newsholme, E.A., 1975, Comparative aspects of fuel utilization and metabolism by muscle, in: *Insect Muscle* (P.N.R. Usherwood, ed.), pp. 405–500, Academic Press, New York.

D'Costa, M.A., and Birt, L.M., 1966, Changes in the lipid content during the metamorphosis of the blowfly, *Lucilia*, *J. Insect Physiol.* **12:**1377.
D'Costa, M.A., and Birt, L.M., 1969, Mitochondrial oxidations of fatty acids in the blowfly, *Lucilia*, *J. Insect Physiol.* **15:**1959.
Demyanovski, S.Y., and Zubova, V.A., 1957, Fats in the organs of the oak silkworm, *Biokhimiya* **21:**698.
Domroese, K.A., and Gilbert, L.I., 1964, The role of lipid in adult development and flight muscle metabolism in *Hyalophora cecropia*, *J. Exp. Biol.* **41:**573.
Downer, R.G.H., 1972, Interspecificity of lipid-regulating factors from insect corpus cardiacum, *Can. J. Zool.* **50:**63.
Downer, R.G.H., 1978, Functional role of lipids in insects, in: *Biochemistry of Insects* (M. Rockstein, ed.), pp. 57–92, Academic Press, New York.
Downer, R.G.H., and Matthews, J.R., 1976, Patterns of lipid distribution and utilisation in insects, *Am. Zool.* **16:**733.
Downer, R.G.H., and Steele, J.E., 1973, Haemolymph lipase activity in the American cockroach, *Periplaneta americana*, *J. Insect Physiol.* **19:**523.
Edwards, G.E., and Ruska, H., 1955, The function and metabolism of certain insect muscles in relation to their structure, *Q. J. Microsc. Sci.* **96:**151.
Edwards, Y.H., Chase, J.F.A., Edwards, M.R., and Tubbs, P.K., 1974, Carnitine acetyltransferase: The question of multiple forms, *Eur. J. Biochem.* **46:**209.
Engelmann, F., 1970, *The Physiology of Insect Reproduction*, Pergamon Press, Oxford.
Engelmann, F., 1979, Insect vitellogenin: Identification, biosynthesis, and role in vitellogenesis, *Adv. Insect Physiol.* **14:**49.
Fast, P.G., 1964, Insect lipids: A review, *Mem. Entomol. Soc. Can.* **37:**1.
Fast, P.G., 1970, Insect lipids, in: *Progress in the Chemistry of Fats and Other Lipids*, Volume XI (R.T. Holman, ed.), Part 2, pp. 181–242, Pergamon Press, Oxford.
Ford, W.C.L., and Candy, D.J., 1972, The regulation of glycolysis in perfused locust flight muscle, *Biochem. J.* **130:**1101.
Fritz, I.B., 1955, Effects of muscle extracts on the oxidation of palmitic acid by liver slices and homogenates, *Acta Physiol. Scand.* **34:**367.
Fritz, I.B., and Marquis, N.R., 1965, The role of acylcarnitine esters and carnitine palmityltransferase in the transport of fatty acyl groups across mitochondrial membranes, *Proc. Natl. Acad. Sci. U.S.A.* **54:**1226.
Fukuda, S., and Takeuchi, S., 1967, Studies on the diapause factor-producing cells in the suboesophageal ganglion of the silkworm, *Bombyx mori*, *Embryologia* **9:**333.
Fulton, R.A., and Romney, V.E., 1940, The chloroform soluble components of beet leafhoppers as an indication of the distance they move in the spring, *J. Agric. Res. (Washington, D.C.)* **161:**737.
Gäde, G., and Holwerda, D.A., 1976, Involvement of adenosine-3′:5′-cyclic monophosphate in lipid mobilization in *Locusta migratoria*, *Insect Biochem.* **6:**535.
Gellissen, G., and Emmerich, H., 1977, Juvenilhormon- und Diglyceridbindende Proteine in der Hämolymphe von *Locusta migratoria*, *Verh. Dtsch Zool. Ges.* **71:**321.
Gellissen, G., and Emmerich, H., 1980, Purification and properties of a diglyceride-binding lipoprotein (LP I) of the hemolymph of adult male *Locusta migratoria*, *J. Comp. Physiol.* **136:**1.
George, J.C., and Talesara, C.L., 1961, The succinic dehydrogenase levels of the pectoral muscles of a few representative types of birds and a bat in relation to the fibre diameter, muscle weight and body weight, *Comp. Biochem. Physiol.* **3:**267.
Gewecke, M., 1972, Antennen und Stirn-Scheitelhaare von *Locusta migratoria* L. als Luftströmungs-Sinnesorgane bei der Flugsteuerung, *J. Comp. Physiol.* **80:**57.

Gewecke, M., 1975, The influence of the air-current sense organs on the flight behaviour of *Locusta migratoria, J. Comp. Physiol.* **103**:79.

Gilbert, L.I., 1967a, Lipid metabolism and function in insects, *Adv. Insect Physiol.* **4**:69.

Gilbert, L.I., 1967b, Changes in lipid content during the reproductive cycle of *Leucophaea maderae* and effects of the juvenile hormone on lipid metabolism *in vitro, Comp. Biochem. Physiol.* **21**:237.

Gilbert, L.I., and Chino, H., 1974, Transport of lipids in insects, *J. Lipid Res.* **15**:439.

Gilbert, L.I., and King, D.S., 1973, Physiology of growth and development: Endocrine aspects, in: *The Physiology of Insecta*, Volume I (M. Rockstein, ed.), pp. 249–370, Academic Press, New York.

Gilbert, L.I., and Schneiderman, H.A., 1961a, The content of juvenile hormone and lipid in Lepidoptera: Sexual differences and developmental changes, *Gen. Comp. Endocrinol.* **1**:453.

Gilbert, L.I., and Schneiderman, H.A., 1961b, Some biochemical aspects of insect metamorphosis, *Am. Zool.* **1**:11.

Gilbert, L.I., Chino, H., and Domroese, K.A., 1965, Lipolytic activity of insect tissues and its significance in lipid transport, *J. Insect Physiol.* **11**:1057.

Goldsworthy, G.J., 1969, Hyperglycaemic factors from the corpus cardiacum of *Locusta migratoria, J. Insect Physiol.* **15**:2131.

Goldsworthy, G.J., 1976, Hormones and flight in the locust, in: *Perspectives in Experimental Biology*, Volume I (P.S. Davis, ed.), pp. 167–177, Pergamon Press, Oxford.

Goldsworthy, G.J., and Coupland, A.J., 1974, The influence of the corpora cardiaca and substrate availability on flight speed and wing beat frequency in *Locusta, J. Comp. Physiol.* **89**:359.

Goldsworthy, G.J., and Mordue, W., 1974, Neurosecretory hormones in insects, *J. Endocrinol.* **60**:529.

Goldsworthy, G.J., Mordue, W., and Guthkelch, J., 1972a, Studies on insect adipokinetic hormones, *Gen. Comp. Endocrinol.* **18**:545.

Goldsworthy, G.J., Johnson, R.A., and Mordue, W., 1972b, *In vivo* studies on the release of hormones from the corpora cardiaca of locusts, *J. Comp. Physiol.* **79**:85.

Goldsworthy, G.J., Jutsum, A.R., and Robinson, N.L., 1979, Substrate utilization and flight speed during tethered flight in the locust, *J. Insect Physiol.* **25**:183.

Hansford, R.G., 1975, The control of tricarboxylate-cycle oxidations in blowfly flight muscle. The oxidized and reduced nicotinamide-adenine dinucleotide content of flight muscle and isolated mitochondria, the adenosine triphosphate and adenosine diphosphate content of mitochondria, and the energy status of the mitochondria during controlled respiration, *Biochem. J.* **146**:537.

Hansford, R.G., and Johnson, R.N., 1976, Some aspects of the oxidation of pyruvate and palmitoylcarnitine by moth (*Manduca sexta*) flight muscle mitochondria, *Comp. Biochem. Physiol.* **55B**:543.

Hill, L., and Goldsworthy, G.J., 1968, Growth, feeding activity, and the utilization of reserves in larvae of *Locusta, J. Insect Physiol.* **14**:1085.

Hill, L., and Izatt, M.E.G., 1974, The relationships between corpora allata and fat body and haemolymph lipids in the adult female desert locust, *J. Insect Physiol.* **20**:2143.

Hochachka, P.W., Neely, J.R., and Driedzic, W.R., 1977, Integration of lipid utilization with Krebs cycle activity in muscle, *Fed. Proc. Fed. Am. Soc. Exp. Biol.* **36**:2009.

Hoffman, A.G.D., and Downer, R.G.H., 1977, Diacylglycerols as major end products of triacylglycerol hydrolysis by tissue lipases of the cockroach, *Am. Zool.* **17**:943.

Hoffman, A.G.D., and Downer, R.G.H., 1979, End product specificity of triacylglycerol lipases from intestine, fat body, muscle and haemolymph of the American cockroach, *Periplaneta americana* L., *Lipids* **14**:893.

Holmes, E.A., and Keeley, L.L., 1975, Mitochondrial development in the flight muscles of the moth, *Heliothis virescens, Insect Biochem.* **5**:15.

Holwerda, D.A., Van Doorn, J., and Beenakkers, A.M.T., 1977, Characterization of the adipokinetic and hyperglycaemic substances from the locust corpus cardiacum, *Insect Biochem.* **7**:151.

Houben, N.M.D., 1976, Regulatie van substraattransport tijdens vlucht in de sprinkhaan, *Locusta migratoria*, Ph.D. Thesis, University of Utrecht, Utrecht.

Johnson, C.G., 1963, Physiological factors in insect migration by flight, *Nature (London)* **198**:423.

Johnson, C.G., 1969, *Migration and Dispersal of Insects by Flight*, Methuen, London.

Johnson, R.N., and Hansford, R.G., 1975, The control of tricarboxylate-cycle oxidations in blowfly flight muscle. The steady-state concentrations of citrate, isocitrate, 2-oxoglutarate and malate in flight muscle and isolated mitochondria, *Biochem. J.* **146**:527.

Jongbloed, J., and Wiersma, C.A.G., 1934, Der Stoffwechsel der Honigbiene während des Fluges, *Z. Vgl. Physiol.* **21**:519.

Jutsum, A.R., and Goldsworthy, G.J., 1976, Fuels for flight in *Locusta, J. Insect Physiol.* **22**:243.

Kallapur, V.L., and George, C.J., 1973, Fatty acid oxidation by the flight muscles of the dragonfly, *Pantala flavescens, J. Insect Physiol.* **19**:1035.

Karuhize, G.R., 1972, Utilization of fat reserve substances by *Homorocoryphus* (Orthoptera: Tettigoniidae) during flight, *Comp. Biochem. Physiol.* **43**:563.

Kinsella, J.E., and Smyth, T., 1966, Lipid metabolism of *Periplaneta americana* L. during embryogenesis, *Comp. Biochem. Physiol.* **17**:237.

Kleinow, W., Sebald, W., Neupert, W., and Bücher, T., 1970, Formation of mitochondria of *Locusta migratoria* flight muscles, in: *Autonomy and Biogenesis of Mitochondria and Chloroplasts* (N.K. Boardman, A.W. Linnane, and R.M. Smillie, eds.), pp. 140–151, North-Holland Publ. Co., Amsterdam.

Klingenberg, M., and Bücher, T., 1960, Biological oxidations, *Annu. Rev. Biochem.* **29**:669.

Kozhantshikov, I.W., 1938, Carbohydrate and fat metabolism in adult Lepidoptera, *Bull. Entomol. Res.* **29**:103.

Krogh, A., and Weis-Fogh, T., 1951, The respiratory exchange of the desert locust (*Schistocerca gregaria*) before, during and after flight, *J. Exp. Biol.* **28**:344.

Krogh, I.M., and Normann, I.C., 1977, The corpus cardiacum neurosecretory cells of *Schistocerca gregaria*. Electron microscopy of resting and secreting cells, *Acta Zool. (Stockholm)* **58**:69.

Kutsch, W., and Gewecke, M., 1979, Development of flight behaviour in maturing adults of *Locusta migratoria*. II. Aerodynamic parameters, *J. Insect Physiol.* **25**:299.

Lambremont, E.N., and Graves, J.B., 1969, Incorporation of acetate-1-^{14}C into neutral lipids and phospholipids during late developmental stages of *Heliothis zea, Comp. Biochem. Physiol.* **30**:347.

Lambremont, E.N., Blum, M.S., and Schrader, R.M., 1964, Storage and fatty acid composition of triglycerides during adult diapause of the boll weevil, *Ann. Entomol. Soc. Am.* **57**:526.

Lee, S.S., and Goldsworthy, G.J., 1975, Allatectomy and flight performance in male *Locusta migratoria, J. Comp. Physiol.* **100**:351.

Lee, S.S., and Goldsworthy, G.J., 1976, The effect of allatectomy and ovariectomy on flight performance in female *Locusta migratoria migratorioides* (R&F), *Acrida* **5**:169.

Levenbook, L., and Williams, C.M., 1956, Mitochondria in the flight muscles of insects. III. Mitochondrial cytochrome c in relation to aging and wing-beat frequency of flies, *J. Gen. Physiol.* **39**:497.

Levinson, Z.H., and Silverman, P.H., 1954, Studies on the lipids of *Musca vicina* (Macq.) during growth and metamorphosis, *Biochem. J.* **58**:294.

Lipsitz, E.Y., and McFarlane, J.E., 1970, Total lipid and phospholipid during the life cycle of the house criquet, *Acheta domesticus* (L.), *Comp. Biochem. Physiol.* **34**:699.

Lipsitz, E.Y., and McFarlane, J.E., 1971, Analysis of lipid during the life cycle of the house criquet, *Acheta domesticus, Insect Biochem.* **1**:446.

Lok, C.M., and Van der Horst, D.J., 1980, Chiral 1,2-diacylglycerols in the haemolymph of the locust, *Locusta migratoria, Biochim. Biophys. Acta* **618**:80.

Ludwig, D., and Ramazzotto, L.J., 1965, Energy sources during embryogenesis of the yellow mealworm, *Tenebrio molitor, Ann. Entomol. Soc. Am.* **58**:543.

Ludwig, D., Crowe, P.A., and Hassemer, M.M., 1964, Free fat and glycogen during the metamorphosis of *Musca domestica* L., *J. N.Y. Entomol. Soc.* **72**:23.

Madariaga, M.A., Municio, A.M., and Ribera, A., 1970a, Biochemistry of the development of the insect *Dacus oleae*: Evolution of fatty acid composition of different lipid classes, *Comp. Biochem. Physiol.* **35**:57.

Madariaga, M.A., Municio, A.M., and Ribera, A., 1970b, Biochemistry of the development of the insect *Ceratitis capitata:* Evolution of fatty acid composition of different lipid classes, *Comp. Biochem. Physiol.* **36**:271.

Martin, J.S., 1969, Lipid composition of fat body and its contribution to the maturing oocytes in *Pyrrhocoris apterus, J. Insect Physiol.* **15**:1025.

Mauldin, J.K., Lambremont, E.N., and Graves, J.B., 1971, Principal lipid classes and fatty acids synthesized during growth and development of the beetle *Lyctus planicollis, Insect Biochem.* **1**:316.

Mayer, R.J., and Candy, D.J., 1967, Changes in haemolymph lipoproteins during locust flight, *Nature* (*London*) **215**:987.

Mayer, R.J., and Candy, D.J., 1969a, Changes in energy reserves during flight of the desert locust, *Schistocerca gregaria, Comp. Biochem. Physiol.* **31**:409.

Mayer, R.J., and Candy, D.J., 1969b, Control of haemolymph lipid concentration during locust flight: An adipokinetic hormone from the corpora cardiaca, *J. Insect Physiol.* **15**:611.

Meyer, H., Preiss, B., and Bauer, S., 1960, The oxidation of fatty acids by a particulate fraction from desert-locust (*Schistocerca gregaria*) thorax tissues, *Biochem. J.* **76**:27.

Mwangi, R.W., and Goldsworthy, G.J., 1977, Diglyceride-transporting lipoproteins in *Locusta, J. Comp. Physiol.* **114**:177.

Nelson, D.R., Terranova, A.C., and Sukkestad, D.R., 1967, Fatty acid composition of the glyceride and free fatty acid fractions of the fat body and hemolymph of the cockroach, *Periplaneta americana* (L.), *Comp. Biochem. Physiol.* **20**:907.

Newsholme, E.A., and Randle, P.J., 1964, Effects of fatty acids, ketone bodies and pyruvate, and of alloxan-diabetes, starvation, hypophysectomy and adrenalectomy, on the concentration of hexose phosphates, nucleotides and inorganic phosphate in perfused rat heart, *Biochem. J.* **93**:641.

Newsholme, E.A., and Taylor, K., 1969, Glycerol kinase activities in muscles from vertebrates and invertebrates, *Biochem. J.* **112**:465.

Newsholme, E.A., Randle, P.J., and Manchester, K.L., 1962, Effects of long chain FFA on glucose uptake, *Nature* (*London*) **193**:270.

Niemierko, S., Wlodawer, P., and Wojtczak, A.F., 1956, Lipid and phosphorus metabolism during growth of the silkworm (*Bombyx mori* L.), *Acta Biol. Exp.* (*Warsaw*) **17**:255.

Nwanze, K.F., Maskarinee, J.K., and Hopkins, T.L., 1976, Lipid composition of the normal and flight forms of adult cowpea weevils, *Callosobruchus maculatus, J. Insect Physiol.* **22**:897.

Odhiambo, T.R., 1966a, The metabolic effects of the corpus allatum hormone in the male desert locust. I. Lipid metabolism, *J. Exp. Biol.* **45**:45.
Odhiambo, T.R., 1966b, The metabolic effects of the corpus allatum hormone in the male desert locust. II. Spontaneous locomotor activity, *J. Exp. Biol.* **45**:51.
Pan, M.L., and Wallace, R.A., 1974, *Cecropia* vitellogenin: Isolation and characterization, *Am. Zool.* **14**:1239.
Pearincott, J.V., 1960, Changes in the lipid content during growth and metamorphosis of the housefly, *Musca domestica* Linnaeus, *J. Cell. Comp. Physiol.* **55**:167.
Peled, Y., and Tietz, A., 1975, Isolation and properties of a lipoprotein from the haemolymph of the locust, *Locusta migratoria*, *Insect Biochem.* **5**:61.
Pette, D., 1965, Plan and Muster im zellulären Stoffwechsel, *Naturwissenschaften* **52**:597.
Pette, D., 1966, Mitochondrial enzyme activities, in: *Regulation of Metabolic Processes in Mitochondria*, Volume 7 (J.M. Tager, S. Papa, E. Quagliriello, and E.C. Slater, eds.), pp. 28–50, Elsevier, Amsterdam.
Pettit, F.H., Pelley, J.W., and Reed, L.J., 1975, Regulation of pyruvate dehydrogenase kinase and phosphatase by acetyl-CoA/CoA and NADH/NAD ratios, *Biochem. Biophys. Res. Commun.* **65**:575.
Pfeiffer, I.W., 1945, Effect of the corpora allata on the metabolism of adult female grasshopper, *J. Exp. Zool.* **99**:183.
Pitts, C.W., and Hopkins, T.L., 1965, Lipid composition of hibernating face flies, *Proc. North Cent. Branch Entomol. Soc. Am.* **20**:72.
Poels, C.L.M., and Beenakkers, A.M.T., 1969, The effect of corpus allatum implantation on the development of flight muscles and fat body in *Locusta migratoria*, *Entomol. Exp. Appl.* **12**:312.
Rademakers, L.H.P.M., and Beenakkers, A.M.T., 1977, Changes in the secretory activity of the glandular lobe of the corpus cardiacum of *Locusta migratoria* induced by flight, *Cell Tissue Res.* **180**:155.
Ramsay, R.R., and Tubbs, P.K., 1975, The mechanism of fatty acid uptake by heart mitochondria: An acylcarnitine-carnitine exchange, *FEBS Lett.* **54**:21.
Rankin, M.A., 1974, The hormonal control of flight in the milkweed bug, *Oncopeltus fasciatus*, in: *Experimental Analysis of Insect Behaviour* (L. Barton Browne, ed.), pp. 317–328, Springer, Berlin.
Rees, K.R., 1954, Aerobic metabolism of the muscle of *Locusta migratoria*, *Biochem. J.* **58**:196.
Robinson, N.L., and Goldsworthy, G.J., 1974, The effects of locust adipokinetic hormone on flight muscle metabolism *in vivo* and *in vitro*, *J. Comp. Physiol.* **89**:369.
Robinson, N.L., and Goldsworthy, G.J., 1976, Adipokinetic hormone and flight metabolism in the locust, *J. Insect Physiol.* **22**:1559.
Robinson, N.L., and Goldsworthy, G.J., 1977, A possible site of action for adipokinetic hormone on the flight muscle of locusts, *J. Insect Physiol.* **23**:153.
Rothstein, F., 1952, Biochemical changes during the embryonic development of the Japanese beetle (*Popillia japonica* Newman), *Physiol. Zool.* **25**:171.
Rudolfs, W., 1926, Studies on chemical changes during the life cycle of the tent caterpillar (*Malacosoma americana* Fab.). I. Moisture and fat, *J. N.Y. Entomol. Soc.* **34**:249.
Sacktor, B., 1965, Energetics and respiratory metabolism of muscular contraction, in: *The Physiology of Insecta*, Volume 2 (M. Rockstein, ed.), pp. 483–580, Academic Press, New York.
Sacktor, B., 1970, Regulation of intermediary metabolism, with special reference to the control mechanisms in insect flight muscle, *Adv. Insect Physiol.* **7**:267.
Sacktor, B., 1975, Utilization of fuels by muscle, in: *Insect Biochemistry and Function* (D.J. Candy and B.A. Kilby, eds.), pp. 1–81, Chapman and Hall, London.

Siakotos, A.N., 1960, The conjugated plasma proteins of the American cockroach. I. The normal state, *J. Gen. Physiol.* **43**:999.

Smith, L.C., Pownall, H.J., and Gotto, A.M., Jr., 1978, The plasma lipoproteins: Structure and metabolism, *Annu. Rev. Biochem.* **47**:751.

Spencer, I.M., and Candy, D.J., 1976, Hormonal control of diacyl glycerol mobilization from fat body of the desert locust, *Schistocerca gregaria, Insect Biochem.* **6**:289.

Staudte, H.W., and Pette, D., 1972, Correlations between enzymes of energy–supplying metabolism as a basic pattern of organization in muscle, *Comp. Biochem. Physiol.* **41B**:533.

Steele, J.E., 1976, Hormonal control of metabolism in insects, *Adv. Insect Physiol.* **12**:239.

Stephen, W.F., and Gilbert, L.I., 1970, Alterations in fatty acid composition during the metamorphosis of *Hyalophora cecropia*: Correlations with juvenile hormone titre, *J. Insect Physiol.* **16**:851.

Stevenson, E., 1966, Rapid oxidation of palmitate with concomitant phosphorylation of adenosine 5′-diphosphate by moth flight-muscle mitochondria, *Biochim. Biophys. Acta* **128**:29.

Stevenson, E., 1968, The carnitine-independent oxidation of palmitate plus malate by moth flight–muscle mitochondria, *Biochem. J.* **110**:105.

Stevenson, E., 1969, Monoglyceride lipase in moth flight muscle, *J. Insect Physiol.* **15**:1537.

Stevenson, E., 1972, Haemolymph lipids and fat body lipases of the southern armyworm moth, *J. Insect Physiol.* **18**:1751.

Stone, J.V., and Mordue, W., 1979, Isolation of granules containing adipokinetic hormone from locust corpora cardiaca by differential centrifugation, *Gen. Comp. Endocrinol.* **39**:543.

Stone, J.V., Mordue, W., Batley, K.E., and Morris, H.R., 1976, Structure of locust adipokinetic hormone, a neurohormone that regulates lipid utilisation during flight, *Nature (London)* **263**:207.

Strong, F.E., 1964, Lipid composition of the egg from an aphid, *Nature (London)* **202**:622.

Strong, L., 1968a, The effect of enforced locomotor activity on lipid content of allatectomized males of *Locusta migratoria migratorioides, J. Exp. Biol.* **48**:625.

Strong, L., 1968b, Locomotor activity, sexual behaviour, and corpus allatum hormone in males of *Locusta, J. Insect Physiol.* **14**:1685.

Thomas, K.K., and Gilbert, L.I., 1968, Isolation and characterization of the hemolymph lipoproteins of the American silkmoth, *Hyalophora cecropia, Arch. Biochem. Biophys.* **127**:512.

Thomas, K.K., and Gilbert, L.I., 1969, The hemolymph lipoproteins of the silkmoth *Hyalophora gloveri*: Studies on lipid composition, origin and function, *Physiol. Chem. Phys.* **1**:293.

Thomsen, E., 1952, Functional significance of the neurosecretory brain cells and the corpus cardiacum in the female blowfly, *Calliphora erythrocephala* Meig., *J. Exp. Biol.* **29**:137.

Tietz, A., 1962, Fat transport in the locust, *J. Lipid Res.* **3**:421.

Tietz, A., 1967, Fat transport in the locust: The role of diglycerides, *Eur, J. Biochem.* **2**:236.

Tietz, A., and Weintraub, H., 1978, Hydrolysis of glycerides by lipases of the fat body of the locust, *Locusta migratoria, Insect Biochem.* **8**:11.

Tietz, A., and Weintraub, H., 1980, The stereospecific structure of haemolymph and fatbody 1,2-diacylglycerol from *Locusta migratoria, Insect Biochem.* **10**:61.

Tietz, A., Weintraub, H., and Peled, Y., 1975, Utilization of 2-acyl-*sn*-glycerol by locust fat body microsomes. Specificity of the acyltransferase system, *Biochim. Biophys. Acta* **388**:165.

Valder, S.M., Hopkins, T.L., and Valder, S.A., 1969, Diapause induction and changes in

lipid composition in diapausing and reproducing faceflies, *Musca autumnalis, J. Insect Physiol.* **15:**1199.
Van den Hondel-Franken, M.A.M., and Flight, W.F.G., 1980, Tracheolization and the effects of implantation of corpora allata on the invagination of tracheoblasts into the developing flight muscle fibers of *Locusta migratoria, Gen. Comp. Endocrinol.* (in press).
Van den Hondel-Franken, M.A.M., Van den Broek, A.T.M., and Beenakkers, A.M.T., 1980, Flight muscle development in *Locusta migratoria*: Effects of implantation of corpora allata on the attainment of metabolic enzyme activities, *Gen. Comp. Endocrinol.* **41:**477.
Van der Horst, D.J., Baljet, A.M.C., Beenakkers, A.M.T., and Van Handel, E., 1978a, Turnover of locust haemolymph diglycerides during flight and rest, *Insect Biochem.* **8:**369.
Van der Horst, D.J., Van Doorn, J.M., and Beenakkers, A.M.T., 1978b, Dynamics in the haemolymph trehalose pool during flight of the locust, *Locusta migratoria, Insect Biochem.* **8:**413.
Van der Horst, D.J., Van Doorn, J.M., and Beenakkers, A.M.T., 1979, Effects of the adipokinetic hormone on the release and turnover of haemolymph diglycerides and on the formation of the diglyceride-transporting lipoprotein system during flight. *Insect Biochem.* **9:**627.
Van der Horst, D.J., Houben, N.M.D., and Beenakkers, A.M.T., 1980, Dynamics of energy substrates in the haemolymph of *Locusta migratoria* during flight, *J. Insect Physiol.* **26:**441.
Van Handel, E., 1974, Lipid utilization during sustained flight of moths, *J. Insect Physiol.* **20:**2329.
Van Handel, E., and Nayar, J.K., 1972, Turnover of diglycerides during flight and rest in the moth *Spodoptera frugiperda, Insect Biochem.* **2:**8.
Van Marrewijk, W.J.A., and Beenakkers, A.M.T., 1979, Regulation of substrate mobilization of flight in locusts, *J. Endocrinol.* **80:**67P.
Van Marrewijk, W.J.A., Van den Broek, A.T.M., and Beenakkers, A.M.T., 1980a, Regulation of glycogenolysis in the locust fat body during flight, *Insect Biochem.*, **10:**675.
Van Marrewijk, W.J.A., Schrikker, A.E.M., and Beenakkers, A.M.T., 1980b, Contents of nucleic and amino acids and rate of protein synthesis in developing flight muscles of *Locusta migratoria, Comp. Biochem. Physiol.* **65B:**251.
Vogell, W., Bishai, F.R., Bücher, T., Klingenberg, M., Pette, D., and Zebe, E., 1959, Über strukturelle und enzymatische Muster in Muskeln von *Locusta migratoria, Biochem. Z.* **332:**81.
Vogt, M., 1949, Fettkörper und Önocyten der *Drosophila* nach Extirpation der adulten Ringdrüse, *Z. Zellforsch. Mikrosk. Anat.* **34:**160.
Vroman, H.E., Kaplanis, J.N., and Robbins, W.E., 1965, Effect of allatectomy on lipid biosynthesis and turnover in the female American cockroach, *Periplaneta americana* (L.), *J. Insect Physiol.* **11:**897.
Wajc, E., and Pener, M.P., 1971, The effect of the corpora allata on the flight activity of the male African migratory locust, *Locusta migratoria migratorioides* (R&F), *Gen. Comp. Endocrinol.* **17:**327.
Walker, P.R., and Bailey, E., 1969, A comparison of the properties of the phosphofructokinases of the fat body and flight muscles of the adult male desert locust, *Biochem. J.* **111:**365.
Walker, P.R., and Bailey, E., 1971, Effect of allatectomy on fat body lipogenic enzymes of the male desert locust during adult development, *J. Insect Physiol.* **17:**1359.

Walker, P.R., Hill, L., and Bailey, E., 1970, Feeding activity, respiration, and lipid and carbohydrate content of the male desert locust during adult development, *J. Insect Physiol.* **16:**1001.

Weeda, E., Koopmanschap, A.B., de Kort, C.A.D., and Beenakkers, A.M.T., 1980, Proline synthesis in fat body of *Leptinotarsa decemlineata, Insect Biochem.* **10:**631.

Weis-Fogh, T., 1952, Fat combustion and metabolic rate of flying locusts, *Philos. Trans. R. Soc. London, Ser. B* **237:**1.

Weis-Fogh, T., 1967, Metabolism and weight economy in migrating animals, particularly birds and insects, in: *Insect Physiology* (J.W.L. Beament, and J.E. Treherne, eds.), pp. 143–159, Oliver and Boyd, Edinburgh.

Wigglesworth, V.B., 1949, The utilization of reserve substances in *Drosophila* during flight, *J. Exp. Biol.* **26:**150.

Williams, C.B., 1958, *Insect Migration*, Collins, London.

Wimer, L.T., and Lumb, R.H., 1967, Lipid composition of the developing larval fat body of *Phormia regina, J. Insect Physiol.* **13:**889.

Wlodawer, P., and Lagwińska, E., 1967, Uptake and release of lipids by the isolated fat body of the wax moth larva, *J. Insect Physiol.* **13:**319.

Wlodawer, P., Lagwińska, E., and Barańska, J., 1966, Esterification of fatty acids in the wax moth haemolymph and its possible role in lipid transport, *J. Insect Physiol.* **12:**547.

Wood, R., Harlow, R.D., and Lambremont, E.N., 1969, GLC analysis of *Heliothis virescens* triglycerides at various metamorphic stages, *Lipids* **4:**159.

Worm, R.A.A., and Beenakkers, A.M.T., 1980, Regulation of substrate utilization in the flight muscle of the locust, *Locusta migratoria*, during flight, *Insect Biochem.* **10:**53.

Worm, R.A.A., Luytjes, W., and Beenakkers, A.M.T., 1980, Regulatory properties of changes in the contents of coenzyme A, carnitine and their derivatives in flight muscle metabolism of *Locusta migratoria, Insect Biochem.* **10:**403.

Wyatt, G.R., and Pan, M.L., 1978, Insect plasma proteins, *Annu. Rev. Biochem.* **47:**779.

Yurkiewicz, W.J., and Oelsner, J., 1969, Neutral lipid metabolism during embryonic development of the Indian meal-moth, *Plodia interpunctella* (Hübner), *Comp. Biochem. Physiol.* **28:**955.

Zebe, E., 1953, Über den respiratorischen Quotienten der Lepidopteren, *Naturwissenschaften* **40:**298.

Zebe, E., 1954, Über den Stoffwechsel der Lepidopteren, *Z. Vgl. Physiol.* **36:**290.

Zebe, E., 1959, Die Verteilung von Enzymen des Fettstoffwechsels im Heuschreckenkörper, *Verh. Dtsch. Zool. Ges.* **31:**309.

Zebe, E., 1960, Condensing Enzyme und β-keto-acyl thiolase in verschiedenen Muskeln, *Biochem. Z.* **332:**328.

4

The Role of Carbohydrate Metabolism in Physiological Function

J.E. Steele

1. Introduction

The purpose of this review is to consider the special role of carbohydrate metabolism in a selected group of physiological mechanisms. It is, to use a now largely discarded term, a study in physiological chemistry in which the relationship between physiological work performed by a tissue and its supporting carbohydrate metabolism is stressed. The reader who is solely interested in the details of the biochemical reactions in which carbohydrates participate has already a number of authoritative reviews available.

The relationship between physiological work and carbohydrate is a particularly important one because carbohydrate is a major source of energy. As a general principle, it is well to keep in mind that energy required for physiological work is provided by energy reserves that are oxidized at rates prescribed by the amount of work being done. In other words, physiological work and metabolism are coupled in such a way as to make the most efficient use of the energy reserves. The great difference in activity between unstimulated and stimulated states in many insect systems indicates their potential importance for the study of coupling mechanisms. The scant attention that this problem has received in insects is not justified in the light of its potential contribution.

J.E. Steele • Department of Zoology, University of Western Ontario, London, Ontario, Canada.

2. Synthesis of Hemolymph Trehalose

Insects are characterized by the presence of the nonreducing disaccharide: trehalose (α-D-glucopyranosyl-α-D-glucopyranoside) in the hemolymph. A second unique feature is the very high concentration of trehalose in the blood compared with that in vertebrates and many other invertebrates. The size of this carbohydrate pool in insects attests to its importance in at least two physiological mechanisms. Development in insects is characterized by the periodic shedding and reconstruction of the cuticle. This event is of obvious importance to the insect and has considerable impact on carbohydrate metabolism because chitin is the major component of the cuticle. Although glycogen can be regarded as the ultimate "deep storage" form of reserve carbohydrate for the synthesis of chitin, trehalose is an intermediate reserve, and its magnitude may be of considerable importance in regulating the rate of chitin synthesis. The transition from rest to work in many insect systems, notably in muscle (see below), often involves a tremendous increase in the rate of energy utilization. The large pool of readily available trehalose in the hemolymph ensures that the need for energy can be met.

2.1. The Biosynthetic Pathway

Two important sources of hemolymph trehalose are fat body glycogen and glucose derived from the diet. When [^{14}C]glucose is injected into adult *Schistocerca*, over 90% of the radioactivity recovered from the hemolymph 15 min later is found in trehalose (Treherne, 1958), indicating the extremely efficient nature of the conversion mechanism. Candy and Kilby (1961) showed that glucose is readily converted into trehalose by *Schistocerca* fat body homogenates and described the enzymatic reactions involved. Simultaneously, Clegg and Evans (1961) provided evidence that fat body glycogen could act as a reserve substrate for the synthesis of trehalose. The biochemical pathway illustrating the reactions by which trehalose is synthesized from glycogen or glucose is shown in Fig. 1.

It can be seen that the synthesis of trehalose is similar to that of glycogen insofar as both require UDP-glucose and glucose-6-PO_4. The synthesis of either product will therefore inevitably affect that of the other, and thus the pathway which predominates will be determined by the concentration of intermediates, products, and substrates. Murphy and Wyatt (1965) have provided a detailed analysis of this system in the silkmoth fat body. Trehalose-6-PO_4 synthetase is controlled by various factors, which include the concentration of glucose-6-PO_4, intracellular Mg^{2+}, and trehalose. The reaction is inhibited by trehalose. Glucose

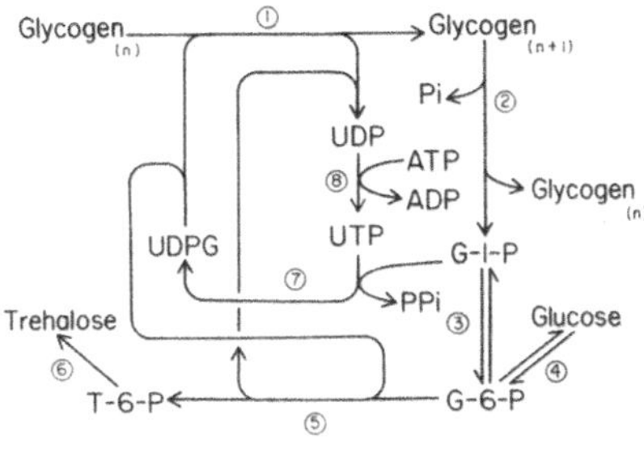

Fig. 1. Pathways for the synthesis of trehalose and glycogen in fat body. The enzymes indicated by the numbers are: (1) glycogen synthetase; (2) glycogen phosphorylase; (3) phosphoglucomutase; (4) hexokinase; (5) trehalose-6-phosphate synthetase; (6) trehalose-6-phosphatase; (7) UDP-glucose pyrophosphorylase; (8) nucleoside diphosphokinase. From Steele (1980). Reproduced with permission of the author and Academic Press, New York.

derived from the diet is converted either into glycogen or trehalose. When the trehalose-6-PO_4 synthetase reaction is not subject to feedback control by trehalose, glucose-6-PO_4 is preferentially converted into trehalose because the K_m for UDPG, the glucosyl donor, is less for trehalose-6-PO_4 synthetase (0.3 mM) than it is for glycogen synthetase (1.6 mM). With increasing synthesis of trehalose, the trehalose-6-PO_4 synthetase reaction is gradually shut off because of feedback inhibition. This causes the concentration of both glucose-6-PO_4 and UDPG to rise. The increasing concentration of glucose-6-PO_4 activates glycogen synthesis because it is an activator for this reaction (K_a = 0.6 mM). The rise in UDPG concentration, which occurs coincident with that of glucose-6-PO_4, allows the reaction rate to increase by virtue of its importance as a donor of glucosyl residues. The importance of trehalose in feedback control of the synthetase reaction has also been demonstrated in *Phormia* fat body, where complete inhibition of trehalose synthesis *in vitro* has been obtained with trehalose concentrations as low as 10% of those occurring naturally (Friedman, 1967). Why should such low concentrations of trehalose cause inhibition *in vitro* but not *in vivo*? The answer to this enigma is unknown. The increased sensitivity of the synthetase reaction *in vitro* may be related to changes in intracellular Mg^{2+} levels, as suggested by the finding in silkmoth pupal fat body that there is a direct correlation between fat body Mg^{2+} concentration and trehalose level (Jungreis and Wyatt, 1972).

2.2. Hormonal Control of Hemolymph Trehalose

2.2.1. *Hemolymph Trehalose Modulation by the Corpora Cardiaca*

A number of studies have shown that the corpora cardiaca can increase the level of trehalose in the hemolymph as a result of their action on the fat body. Not all insects show this response (Chalaye, 1969), even though the corpora cardiaca are known to contain the hyperglycemic principle. Mordue and Goldsworthy (1969) demonstrated that extracts of *Schistocerca* corpora cardiaca elicit a strong hyperglycemic effect when injected into *Periplaneta* but have no effect in the host species. Since extirpation of the corpora cardiaca in *Locusta* actually lowers trehalose levels (Cazal, 1971), it is possible that the glands are effective only when the concentration of trehalose in the hemolymph falls below a certain level; otherwise feedback control of its synthesis prevents the stimulatory effect of the hormone. If the corpus cardiacum extract is injected into decapitated cockroaches the hyperglycemic response is still obtained (Hanaoka and Takahashik 1976), suggesting that the action of the factor is independent of any other factor produced by the remaining cephalic organs. The number of factors present in the neuroendocrine system able to induce a hyperglycemic effect is still in doubt. In this context it is interesting to note that the adipokinetic hormone has been shown to have a hyperglycemic response when injected into cockroaches (Jones *et al.*, 1977).

2.2.2. Regulation of Fat Body Glycogen

The active factor in the corpus cardiacum, by virtue of its ability to stimulate the conversion of glycogen to trehalose, is to a limited extent a regulator of glycogen levels in the fat body. All of the additional trehalose appearing in the hemolymph following the injection of corpus cardiacum extract originates from glycogen in the fat body (Steele, 1963; Hanaoka and Takahashi, 1976). Although the corpora cardiaca do not raise trehalose levels in *Locusta* (Chalaye, 1969), they are potent activators of phosphorylase (Goldsworthy, 1970), and this causes an increase in glycogenolysis (Goldsworthy, 1969). The fact that trehalose in the hemolymph does not increase may reflect the presence of very low levels of glycogen in the fat body, which precludes the synthesis of significant amounts of trehalose. Alternatively, glucose-6-PO_4 arising from glycogen may be channeled into other pathways. The inability of the corpus cardiacum to stimulate glycogenolysis in fed *Phormia* (Friedman, 1967) is undoubtedly due to feedback control of the trehalose synthetase reaction. This conclusion is based on the finding that starvation for 24 hr allows glycogenolysis to be increased by the gland extract (Friedman, 1967). It is not clear how feedback inhibition of the synthetase reaction is translated into an inhibitory effect on glycogenolysis. It would not be surprising to find that this resulted from an accumulation of one or more glycolytic intermediates or possibly UDPG.

2.2.3. Mechanism of Corpus Cardiacum Action

The capacity of corpora cardiaca extracts to activate phosphorylase by converting the b to the a form is well known and is perhaps the only event occuring at the subcellular level which is known in detail. In *Periplaneta*, the activation of phosphorylase is coincident with a doubling of the glucose-6-PO_4 level in fat body (Coulthart and Steele, unpublished observations). This will have a number of consequences that are predictable on the basis of present knowledge. The increase in glucose-6-PO_4 suggests that the rate of synthesis of glucose-1-PO_4 must also increase, since it is the precursor for glucose-6-PO_4 (Fig. 1). This is likely to stimulate trehalose synthesis, since glucose-1-PO_4 is the obligatory glucosyl donor for the synthesis of UDPG. It is possible that the variation in response by different insects to the corpus cardiacum extract merely reflects a difference in the kinetic properties of the various enzymes required for the synthesis of trehalose from glycogen. A comparison of normal hemolymph trehalose levels relative to the ease with which the corpus cardiacum can elevate them might provide some useful data in this matter.

2.2.3a. Activation of glycogen phosphorylase. Regulation of phosphorylase is a complex process requiring the participation of other enzymes, as shown in Fig. 2. The enzyme in fat body is interconvertible between active and inactive forms (Stevenson and Wyatt, 1964). Like mammalian liver phosphorylase, the inactive phosphorylase can be made to appear active by the addition of 5'AMP to the assay medium (Stevenson and Wyatt, 1964). In the cell, the conversion of phosphorylase b to a is by means of a kinase (Wiens and Gilbert, 1967; Yanagawa and Horie, 1978; Ashida and Wyatt, 1979). The two latter studies also show that phosphorylase kinase is activated by ATP and Mg^{2+}. Ca^{2+} further enhances the activation (Ashida and Wyatt, 1979). The requirement for ATP and Mg^{2+} suggests that activation of the kinase may be mediated by phosphorylation of the enzyme protein.

Treatment of *Periplaneta* fat body with corpus cardiacum extract *in vitro* increases the total activity of phosphorylase by 150% (Steele, 1963). These results were confirmed in *Leucophaea* fat body, where it was shown that the gland extract increased the phosphorylase a/b ratio (Wiens

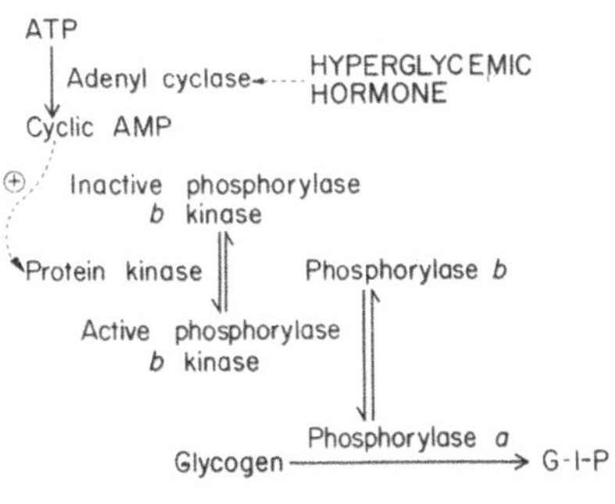

Fig. 2. A scheme illustrating a possible mechanism for the regulation of phosphorylase activity by hyperglycemic hormone. It should be noted that no evidence has yet been obtained to show that phosphorylase kinase is regulated by a cAMP-dependent protein kinase. The scheme also describes the activation of muscle phosphorylase, although in this tissue there may be no involvement of hormones or cAMP. From Steele (1980). Reproduced with permission of the author and Academic Press, New York.

and Gilbert, 1967). The fact that *Locusta* fat body does not increase trehalose synthesis in response to corpus cardiacum extract but does show activation of phosphorylase has already been referred to. It is presumed that activation of phosphorylase in this species occurs by the same mechanism as those that give a hyperglycemic response. This contrasts with the situation in *Hyalophora cecropia*, where not only is there no hyperglycemic response to the extract but also no activation of phosphorylase (Wiens and Gilbert, 1967).

2.2.3b. The function of cAMP. Adenyl cyclase, the function of which is to catalyse the conversion of ATP to cAMP, has been found in the fat body of various insects. These include *Bombyx mori* (Morishima, 1978, 1979), *Hyalophora cecropia* (Filburn and Wyatt, 1976), and *Periplaneta* (Hanaoka and Takahashi, 1977). Morishima (1979) has shown that adenyl cyclase in silkworm pupal fat body is regulated by a Ca^{2+}-dependent regulator protein, the activity of which is determined by Ca^{2+} concentrations in the same range as those in the cell. This finding is of great importance from the viewpoint of hormonal control of metabolism.

The corpora cardiaca increase considerably the concentration of cAMP in the fat body. In many mechanisms, cAMP has been shown to exert its effects through a protein kinase, which in turn phosphorylates a protein such as phosphorylase kinase. There are at least two cAMP-dependent protein kinases in *Periplaneta* fat body (Takahashi and Hanaoka, 1977), thus confirming an essential link in the mechanism of action of cAMP in fat body. A suggested mechanism of action by the corpora cardiaca on phosphorylase activation is shown in Fig. 2. Since the corpora cardiaca can increase the level of cAMP in fat body, the fact that the cyclic nucleotide can activate phosphorylase (Steele, 1964; Ziegler *et al.*, 1979) suggests that it is the intermediary in the activation of phosphorylase by virtue of its action on phosphorylase kinase. This view, however, may not be correct, since partly purified phosphorylase kinase does not appear to be activated by cAMP (Ashida and Wyatt, 1979).

2.2.3c. Calcium dependence of the corpora cardiaca. The hypertrehalosemic effect of the corpora cardiaca on trehalose synthesis in cockroach fat body is absolutely dependent on Ca^{2+} in the extracellular fluid (Steele *et al.*, unpublished observations). The maximum Ca^{2+} concentration required for this effect in *in vitro* experiments was slightly less than that in the hemolymph. Furthermore, Ca^{2+} is a potent activator of phosphorylase kinase in fat body (Ashida and Wyatt, 1979) and may possibly substitute for cAMP in this regard as a second messenger.

2.3. The Source of Energy for Trehalose Synthesis

The synthesis of blood sugar in insects differs from that in vertebrates because UDPG required for the synthetase reaction must be regenerated following each glucosylation step. ATP required for the generation of UDPG from UDP appears to be derived from the oxidation of lipid rather than carbohydrate, although this conclusion may apply only to the hormone stimulated condition.

The respiration of isolated fat body is linearly related to trehalose synthesis, although the correlation is not significant (White and Steele, unpublished observations). This may be because the synthesis of trehalose under resting conditions accounts for only a small part of the ATP being generated. If, on the other hand, the fat body is treated with corpus cardiacum extract there is a significant rise in the rate of respiration, and the correlation with trehalose output by the tissue becomes highly significant (White and Steele, unpublished observations). The increase in respiration is not accompanied by an increase in glycolytic activity, but there is a significant rise in the rate of oxidation of fatty acid (palmitate) and acetate (Wiens and Gilbert, 1965). These results show that cardiacum extract enhances citric acid cycle activity, proving that lipid is the source of energy used for trehalose synthesis. Wiens and Gilbert (1965) have suggested that the active factor acts at two sites. One of these, phosphorylase, is now well established. The other site was not identified but was suggested to be in the glycolytic pathway, where its action reduced the availability of substrate, thus causing oxidation of lipid in the citric acid cycle to be emphasized. Because the concentration of glucose-6-phosphate doubles when isolated fat body is treated with corpus cardiacum extract (Coulthart and Steele, unpublished observations) but no corresponding increase in glycolytic flux occurs (Wiens and Gilbert, 1965), it seems clear that inhibition of one or more enzymatic steps must have occurred. Although cardiacum extract stimulates lipid oxidation, it remains to be shown that lipase activity increases. This appears to be a key problem in explaining the role played by hyperglycemic hormone in the metabolism of carbohydrate and lipid.

3. Carbohydrate as an Energy Source in Flight Muscle

The transition from rest to flight in insects is characterized by a dramatic increase in oxygen consumption. In the blowfly, *Lucilia sericata*, the increase may be as great as 100-fold (Davis and Fraenkel, 1940). Similar increases have also been demonstrated in various species of moths (Zebe, 1954). The fact that excitation–contraction coupling in flight muscle is fundamentally different in each group does not appear to be related to the magnitude of the increment in respiration. It is, however, related to the absolute amount of oxygen consumed. Insects having asynchronous types of muscle, such as the Diptera and Hymenoptera, consume more oxygen during rest and flight than do the Lepidoptera, for example, which have muscles of the synchronous type, which contract at a much lower frequency. The fact that vertebrates do not generally exhibit increases in oxygen consumption greater than 2- to 5-fold as a result of physical activity suggests that the flight muscles of insects must possess certain unique features.

Carbohydrate is an important source of energy in most insects, although its importance varies. Sacktor (1965) has collected data showing that, in general, the Diptera and Hymenoptera have a respiratory quotient (RQ) of 1, indicating that carbohydrate is the sole source of energy. It was long believed that Lepidoptera and Orthoptera obtained their entire energy requirements from lipid. That this view is incorrect was suggested by the finding that homogenized flight muscle of *Hyalophora cecropia* and *Prodenia eridania* readily oxidizes glycogen and trehalose (Gussin and Wyatt, 1965; Stevenson, 1968). The idea that Lepidoptera can derive some of the energy required for flight from carbohydrate, although it may not be a major source *in vivo*, was shown by Van Handel and Nayar (1972) in the moth *Spodoptera*, where oxidation of glucose is greatly increased at the onset of flight. In other insects, such as locusts, there is clear evidence to show that both carbohydrate and lipid are important sources of energy in flight muscle. In these insects, most of the energy is initially obtained from carbohydrate, but lipid subsequently becomes the major source.

A unique feature of oxidative metabolism in the flying insect is that very little, if any, oxygen debt is incurred. This is particularly remarkable in view of the very high rate of metabolism achieved by flight muscle and maintained for long periods (Krogh and Weis-Fogh, 1951). The highly aerobic nature of insect respiration is made possible by their generally small size and efficient oxygen delivery system (Weis-Fogh, 1964).

In the following discussion, regulation of glycolysis and reactions responsible for supplying it with substrate will be examined. The pathway is expressly concerned with the initial steps in the oxidation of carbohydrate. Glycolytic end products are oxidized further in the citric acid cycle, as are metabolic degradation products derived from lipid and amino acids. Because the oxidation of carbohydrate in the citric acid cycle cannot be divorced from that of other substrates, notably lipid, discussion of this particular pathway would be somewhat irrelevant in the present context.

3.1. Glycogen as an Energy Source

Of the two major carbohydrates serving as a source of energy in muscle during flight, glycogen is quantitatively the most important. In the blowfly *Phormia*, the concentration of polysaccharide may reach 15 mg/g

(Clegg and Evans, 1961), and *Periplaneta*, although not a strong flyer, has a similar concentration of glycogen in its thoracic musculature (Steele, 1963; Downer and Parker, 1979). Even *Locusta*, generally considered to obtain most of its energy requirements from lipid, is now known to have glycogen reserves in the flight muscle equalling those in *Phormia* muscle (Worm and Beenakkers, 1980).

There is very little utilization of thoracic muscle glycogen during the first minute or two of flight in the blowfly, but most has been consumed after 15 min (Sacktor and Wormser-Shavit, 1966). A similar situation has been observed in *Periplaneta*, where glycogen decreases from 14.8 mg/g to 0.98 mg/g after 10 min of flight (Downer and Parker, 1979). These two examples clearly illustrate that glycogen in muscle is an important source of energy in the early stages of flight. The locust, also, depends strongly on its glycogen reserves early in flight; the reserves being depleted by 50% during the first 5 min (Worm and Beenakkers, 1980).

3.1.1. Control of Glycogenolysis

The first step in the degradation of glycogen involves a cleavage of glucosyl residues from the outer chains of the glycogen molecule. The enzyme responsible for this is phosphorylase. The reaction is of special significance because it is a rate limiting reaction, and thus the funneling of glucosyl residues into the glycolytic pathway is subject to considerable modulation. In view of the large increases in activity flight muscle is capable of, it is surprising that phosphorylase has received as little attention as it has.

3.1.2. Activation of Phosphorylase

Phosphorylase exists in two interconvertible forms. The more active form is designated phosphorylase a and the less active form phosphorylase b. The relative amount of each is determined by two enzymes. These are phosphorylase kinase which converts the b form to a and phosphorylase phosphatase which catalyzes the conversion in the opposite direction. The molecular weight of both phosphorylase a and b is approximately 100,000 daltons (Childress and Sacktor, 1970). Thus, it appears that activation is achieved through phosphorylation alone rather than phosphorylation and dimerization, as is the case with vertebrate skeletal muscle phosphorylase. The activation and inactivation of phosphorylase by its associated enzymes is shown in Fig. 2.

3.1.2a. Kinetic properties of phosphorylase. At a given time, the actual catalytic activity of phosphorylase is determined by a large number of factors, the effect of which may be altered as the physiological state changes. In considering the activity of phosphorylase, the potential contribution of phosphorylase a and b acting separately must be considered. Childress and Sacktor (1970) have shown that apart from the obvious requirement for inorganic phosphate (Pi) and glycogen as substrates, the relative concentration of each is important. This is because the binding of Pi to the enzyme facilitates the binding of glycogen and *vice versa*. For example, in the study cited it was shown that the K_m for Pi with 0.15 mM glycogen was 47 mM, which decreased to 7.3 mM with 1.7 mM glycogen. The same study showed that the concentration of Pi increased after 15 min of flight from 7.0 to 7.5 mM and thus would tend to enhance the binding of glycogen, as the concentration of substrate decreased with flight. The activity of phosphorylase a is greatly enhanced by AMP, although this nucleotide is not a mandatory cofactor for activity. The stimulatory effect of AMP on phosphorylase activity is of physiological importance. At low concentrations of glycogen, the activity of the enzyme may be increased as much as 10-fold. In situations where the level of glycogen is saturating, the increase in activity due to AMP may still be as great as 2-fold (Childress and Sacktor, 1970). It appears that AMP acts by facilitating the binding of the substrate to the enzyme; the K_m for Pi decreasing from 100 mM to 9 mM as the concentration of AMP rises. Since the concentration of AMP in flight muscle has been shown to rise 2-fold during flight (Childress and Sacktor, 1970), the importance of this activator is readily apparent. In contrast to AMP, ATP is without effect on the activity of phosphorylase a.

Under appropriate *in vitro* conditions, phosphorylase b exhibits strong activity. These conditions include the presence of AMP, which is an essential activator (Childress and Sacktor, 1970). It appears unlikely, however, that suitable conditions involving AMP ever arise *in vivo*. Since the concentration of Pi in muscle during flight is approximately 7.5 mM, it can be shown that the K_m for AMP under these conditions is approximately 1.0 mM (Childress and Sacktor, 1970). This is 10-fold higher than the concentration in the tissue, even after allowing for the 2-fold rise in AMP that occurs after a few minutes of flight. The same authors have shown that ATP at the concentration present in the tissue (7 mM) is strongly inhibitory to phosphorylase b, thus confirming the view that the b form of the enzyme has no significant activity under *in vivo* conditions.

With the onset of flight in the blowfly, glycogen is utilized at the rate of 2.5 μmol of glucosyl residues per min per g thorax (Sacktor and Wormser-Shavit, 1966). Sacktor (1975) has estimated that in order to accommodate this rate of glycogenolysis at least 50% of the phosphorylase must be in the a form. Actual measurements of phosphorylase a/b ratios during rest and flight show that the relative amount of the a form of the enzyme rises from 18% at rest to 70% fifteen seconds after the initiation of flight (Childress and Sacktor, 1970). This level is maintained for the duration of flight. The rate of glycogenolysis can therefore be sustained unless the concentration of Pi or glycogen becomes limiting. As Childress and Sacktor (1970) have shown, the actual concentration of Pi in the muscle is equal to the K_m for phosphorylase a; thus, the activity of the enzyme is sufficient to accommodate the intense rate of glycogenolysis essential for flight activity.

3.1.2b. Role of Ca^{2+} in phosphorylase activation. Within 15 sec after the initiation of flight, the amount of phosphorylase a rises sharply from 18% to 70% (Childress and Sacktor, 1970). The conversion of phosphorylase a to b is catalyzed by phosphorylase kinase. In marked contrast to skeletal muscle phosphorylase kinase, the insect flight muscle enzyme does not appear to require ATP nor is it activated by a protein kinase (Sacktor *et al.*, 1974). The kinase is, however, activated by Ca^{2+} at concentrations anticipated to occur in the cell. The range of concentrations over which Ca^{2+} will activate the enzyme is 10^{-8} M–10^{-6} M (Hansford and Sacktor, 1970; Sacktor *et al.* 1974). The flight muscle enzyme differs from that of rabbit muscle phosphorylase kinase in that it shows about 30% activity in the absence of Ca^{2+}, whereas the vertebrate enzyme is devoid of activity. Furthermore, the flight muscle kinase is less sensitive to Ca^{2+} than the vertebrate enzyme in that the increase in activity produced by Ca^{2+} is much less. It has been suggested that the functional significance of these observations is related to the fact that flight muscle does not possess the well developed and extensive sarcoplasmic reticular system found in vertebrate skeletal muscle. The conclusion is drawn that the normal operation of the oscillatory contractile mechanims does not require the removal of Ca^{2+} from the contractile elements, although the development of tension depends on an increase in concentration of Ca^{2+} in the sarcoplasm.

3.2. Trehalose as an Energy Source

The very high concentration of trehalose in hemolymph, as characterized by the 20 mg/ml value for blowfly hemolymph (Clegg and Evans, 1961), is indicative of its importance as an energy source. This is further emphasized by the finding that trehalose levels, in both the hemocoel and thoracic muscle, are significantly reduced after flight. The effect of flight has been observed in a number of species, including *Locusta* (Houben and Beenakkers, 1975; Jutsum and Goldsworthy, 1976), *Periplaneta* (Polacek and Kubista, 1960), and *Phormia* (Clegg and Evans, 1961; Sacktor and Wormser-Shavit, 1966). The clear relationship between the ability of the thoracic musculature to perform work and the concentration of trehalose in the hemolymph is illustrated in Fig. 3.

The metabolism of trehalose begins with the cleavage of the molecule into two glucose residues. This hydrolytic reaction is catalyzed by the enzyme trehalase. A number of studies have confirmed that two forms of the enzyme, one soluble the other particulate, occur in thoracic muscle. The particulate enzyme, constituting approximately 75% of the total, has been reported to occur in the microsomal fraction of *Hyalophora* (Gussin and Wyatt, 1965) and *Blaberis* (Gilby *et al.*, 1967) muscle. In *Phormia*, *Sarcophaga*, and *Leucophaea* it has been demonstrated to occur in the mitochondria (Zebe and McShan, 1959; Hansen, 1966; Clements *et al.* 1970; Reed and Sacktor, 1971). The soluble enzyme, making up the remainder, is not identical with the enzyme in hemolymph (Friedman and Alexander, 1971).

The utilization of trehalose in blowfly muscle, in contrast to that present in hemolymph, is very rapid during the first minute or two of

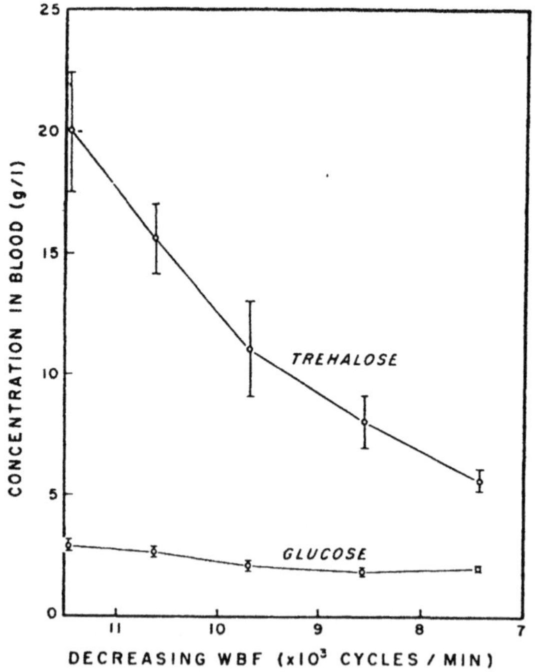

Fig. 3. Relationship between wing-beat frequency (WBF) and the concentration of blood glucose and trehalose. From Clegg and Evans (1961). Reproduced with permission of the author. Copyright 1961 by the American Association for the Advancement of Science.

flight and decreases thereafter at a less rapid but constant rate (Sacktor and Wormser-Shavit, 1966). This suggests that two separate pools of trehalose are being metabolized, the more rapidly utilized pool being that within the cell and the less rapid that present in the hemolymph bathing the tissue. Worm and Beenakkers (1980), on the basis of statistical analysis, have concluded that trehalose levels in the thoracic musculature of locusts do not change during flight. An explanation of these data, however, shows a pattern of utilization not unlike that of the blowfly; a comparison that might be confirmed by use of larger sample sizes.

Because the pattern of change in trehalose with time after the start of flight is similar for the locust and blowfly, it might be anticipated that glucose levels would also show the same pattern in both species. This does not occur. In the blowfly, there is a rapid increase in glucose concentration within seconds, followed by a decline to subnormal levels before returning to the starting concentration (Sacktor and Wormser-Shavit, 1966). By contrast, glucose levels in the locust decline rapidly for the first 5 min of flight and more slowly thereafter (Worm and Beenakkers, 1980), without exhibiting the oscillatory pattern found in *Phormia*. These observations suggest two possibilities. First, the rate at which hemolymph trehalose is made available to the intracellular trehalase is faster in some species than others, and second, the activity of trehalase may be regulated. Since muscle lacks glucose-6-phosphatase, the free glucose present must be derived in large measure from the hydrolysis of trehalose. The constancy of muscle glucose in the blowfly, in spite of declining levels of this sugar in the hemolymph with flight time, suggests regulation of the uptake mechanism. The declining level of trehalose in the tissue also suggests that increased trehalase activity may be part of the mechanism for maintaining glucose levels. Declining levels of glucose in locust muscle suggest that there is no enhancement of trehalase activity in that tissue, since trehalose remains reasonably constant. This fact, however, suggests that the uptake mechanism may be subject to modulation, since the concentration of trehalose in blood decreases to 50% of the normal value during flight (Houben and Beenakkers, 1975; Jutsum and Goldsworthy, 1976).

3.3. Glycolysis

The transition from rest to flight in many insects is accompanied by a 100-fold increase in the metabolic rate. In those insects where carbohydrate is the sole source of energy for flight, this means that a comparable increase in glycolytic flux can be expected. The maximum rate of glycolytic flux is determined by the enzyme having the lowest activity, assuming, of course, that the supply of hexose does not become limiting.

The exception to this rule is glycogenolysis, since glucose-6-phosphate that arises from glycogen can also be derived from trehalose. It is necessary, however, that the sum of the activities of phosphorylase and hexokinase provide sufficient glucose-6-phosphate to saturate phosphoglucoisomerase. There is an interesting difference between the locust and blowfly with respect to regulation of glucose-6-phosphate during rest and flight. In the locust, the concentration of glucose-6-phosphate is doubled shortly after initiation of flight but that of glucose-1-phosphate remains unchanged (Worm and Beenakkers, 1980). Since the level of glucose declines with the onset of flight in the locust (Worm and Beenakkers, 1980), it appears that glucose or trehalose or both are the source of the additional glucose-6-phosphate. In the blowfly, on the other hand, glucose-6-phosphate declines to half the resting value 15 min after the start of flight (Sacktor and Wormser-Shavit, 1966). Because glucose is stabilized at the resting level, it seems that the reduction in glucose-6-phosphate can be accounted for only by the corresponding change in the concentration of glycogen. Since there is an increased flux of hexose through the isomerase reaction with the onset of flight in both locust and blowfly, the fact that fructose-6-phosphate levels remain remarkably constant (Sacktor and Wormser-Shavit, 1966; Worm and Beenakkers, 1980) suggests two ways in which this might occur. If the glucose-6-phosphate level is not saturating (as might be the situation in the resting locust) an increase in concentration would cause the rate of catalysis to rise and thus maintain the level of fructose-6-phosphate. A decrease in the K_m for the enzyme and its substrate would achieve the same effect and might explain the maintenance of fructose-6-phosphate levels in the blowfly when there is a decreasing concentration of glucose-6-phosphate. It is unfortunate that we do not have comparative kinetic data for the isomerase from different species.

A compilation of data on the activity of individual glycolytic enzymes in various species as described in Sacktor (1965) and Crabtree and Newsholme (1975) shows that the rate-limiting enzyme in the glycolytic sequence in flight muscle is phosphofructokinase (PFK). Additional evidence for this view has been provided by Sacktor and Wormser-Shavit (1966), who showed that the beginning of flight in the blowfly coincided with a decrease in the level of total hexosemonophosphate but a rise in that of fructose diphosphate. The fact that individual activities of hexokinase and phosphorylase are less than that of PFK is not significant, since the sum of both is greater than that of PFK. A particularly perplexing problem is the finding that the activity of PFK under what is thought to be optimal conditions falls short of that required to explain the very large increase in the glycolytic rate when flight begins. Like the mammalian enzyme, PFK from flight muscle is inhibited by excess ATP (Grasso and Migliore-

Natalizi, 1968; Walker and Bailey, 1969; Vaughan *et al.*, 1973). During flight, the concentration of ATP in the blowfly falls from 6.9 mM to 6.2 mM (Sacktor and Hurlbut, 1966), but even this concentration is inhibitory. The inhibition caused by ATP can be overcome by AMP, cAMP, and Pi (Walker and Bailey, 1969), so that these factors undoubtedly play an important role *in vivo*. In the blowfly, the concentration of Pi rises slightly when flight begins, whereas that of AMP is doubled (Sacktor and Hurlbut, 1966). Furthermore, the concentration of Pi and NH_4^+, both of which have been shown to increase in locust flight muscle after 3 min of flight (Rowan and Newsholme 1979), are activators of PFK independent of any effects of ATP (Sugden and Newsholme, 1975). It is noteworthy that the insect enzyme, unlike its mammalian counterpart, is inhibited by fructose diphosphate (Walker and Baily, 1969); thus, the difficulty in explaining the activation of PFK is compounded. In this regard, it is interesting to note that fructose diphosphate in both locust and blowfly muscle is significantly lower during flight than at rest (Sacktor and Wormser-Shavit, 1966; Worm and Beenakkers, 1980). In contrast to these findings, Rowan and Newsholme (1979) found fructose diphosphate in locust muscle to be higher 3 min after the beginning of flight than during rest, but these determinations were probably made before steady state conditions were achieved (cf. Worm and Beenakkers, 1980). It seems clear that ATP, AMP, NH_4^+, Pi, and fructose diphosphate all play a role in determining the activity of PFK, although essential elements in the mechanism remain to be described. An interesting possibility suggested by Newsholme (described in Newsholme and Start, 1973) makes use of the fact that interconversion of fructose-6-phosphate and fructose-1,6-diphosphate is by means of a substrate cycle. Whereas the activity of PKF is activated by AMP, that of fructose-1,6-diphosphatase is inhibited. Since the activity of the diphosphatase is much less than that of the kinase, it can be shown that the net effect of AMP on PFK activity is sigmoidal in nature. This makes it possible for relatively small changes in AMP concentration to have proportionately greater effects on the activity of PFK.

The oxidation of glyceraldehyde-3-phosphate produced by the aldolase reaction is of particular importance because it is coupled to the reduction of NAD^+. Under ideal conditions, the activity of glyceraldehyde-3-phosphate dehydrogenase does not limit glycolytic flux, but since NAD^+ is present in catalytic quantities, only the mechanism for the oxidation of NADH, particularly in working flight muscle, is of major importance. In working vertebrate muscle, functioning under hypoxic conditions, this need is met by the lactate dehydrogenase reaction. Insect flight muscle, by contrast, is characterized by a lack of LDH but very high α-glycerophosphate dehydrogenase activity (for comparative data for a number of species see the review by Sacktor, 1975). There is convincing evidence that the α-glycerophosphate dehydrogenase is part of a shuttle system whereby reducing equivalents originating from the oxidation of glyceraldehyde-3-phosphate are transferred from the cytoplasm to the mitochondria (Fig. 4). The shuttle mechanism was proposed in order to explain the production of NAD^+ in flight muscle, because it lacked LDH (Zebe and McShan, 1957; Chefurka, 1958). The mechanism also takes into account the fact that flight muscle actively metabolizes glycerol-3-phosphate and, under anaerobic conditions, accumulates both glycerol-3-phosphate and pyruvate. The shuttle is required because NADH formed in the glyceraldehyde phosphate dehydrogenase (GAP DH) reaction cannot permeate the mitochondrial membrane and thus is unavailable to the electron transport system for oxidation. This is achieved by coupling the oxidation of NADH to the reduction of DHAP to yield glycerol-3-phosphate in the cytoplasmic compartment. The glycerol-3-phosphate diffuses into the mitochondria where it is oxidized to DHAP, and NAD^+ is reduced to NADH. It can be seen in Fig. 4 that the shuttle operates as a cycle, in which catalytic amounts of DHAP are required. This permits the major part of the hexose substrate to be channeled through the glyceraldehyde phosphate dehydrogenase reaction. The special significance of the glycerol-3-phosphate shuttle is that it permits glycolysis to proceed without the accumulation of blind pathway end products such as lactic acid. Pyruvate is therefore continually available to the citric acid cycle, to the extent permitted by glycolytic activity. The

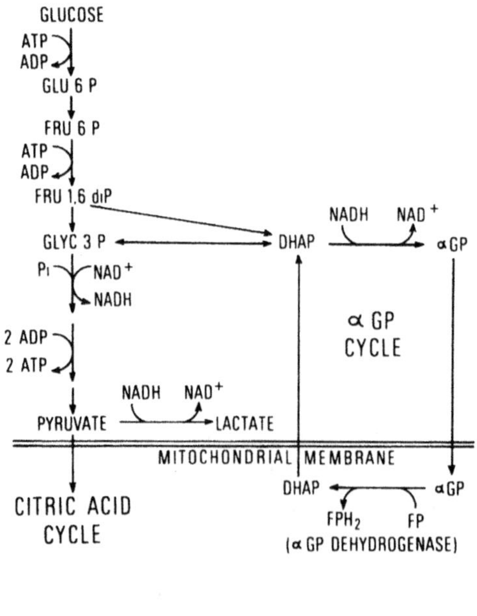

Fig. 4. Schematic diagram of glycolysis and the α-glycerophosphate shuttle. From Sacktor (1970). Reproduced with permission of the author and Academic Press, London.

rate of glycolysis, therefore, depends on the rate at which NAD^+ is supplied by the glycerol-3-phosphate dehydrogenase reaction. Crabtee and Newsholme (1972) have compared the glycerol-3-phosphate dehydrogenase activity in the cytoplasmic and mitochondrial compartments of flight muscle for a number of species and have shown that in insects such as cockroaches, locusts, and flies the cytoplasmic activity often exceeds that in the mitochondria by as much as threefold. In most instances, the mitochondrial activity was approximately double or greater than that of PFK. The explanation for the PFK: glycerol-3-phosphate dehydrogenase ratio of 2:1 is probably that the PFK reaction produces a hexose moiety that is subsequently cleaved to yield two trioses, the metabolism of which is coupled to the reduction of NAD^+ and also to the glycerol-3-phosphate dehydrogenase reaction. The very high cytoplasmic activity of glycerol-3-phosphate dehydrogenase ensures that this step in the shuttle does not become limiting. Thus, it can be seen that the shuttle provides NAD^+ at a rate compatible with the maximum glycolytic rate as determined by the PFK reaction.

There is no evidence that flight muscle contains a phosphatase capable of cleaving the phosphate from glycerol-3-phosphate to yield free glycerol. The absence of such an enzyme ensures that the shuttle mechanism has first call on the glycerol-3-phosphate, thus maintaining shuttle activity at the highest possible level. The reason why the enzyme catalyzing the reaction in the opposite direction is present is not apparent. Newsholme and Taylor (1969) have shown that flight muscles can be characterized by low, medium, or high levels of glycerol kinase activity. It has been suggested that the function of glycerol kinase is to initiate the metabolism of glycerol that accumulates in the muscle of those insects that utilize lipid. This accumulation has been demonstrated by Worm and Beenakkers (1980). The role of the enzyme in those insects utilizing primarily carbohydrate in their flight muscle is unknown. Perhaps there is an extramuscular source of glycerol in those insects.

4. The Biosynthesis of Antifreeze

A remarkable adaptation found in certain insects is the ability to synthesize substances that prevent the formation of ice crystals that otherwise would have a lethal effect. The production of these "antifreeze" molecules is usually associated with a state of diapause. In some insects at least, diapause is induced by a hormone, and thus it seems likely that antifreeze production is under hormonal control. There is also evidence that the mere lowering of the ambient temperature is sufficient to initiate synthesis of the antifreeze. In this section the present state of knowledge of antifreeze synthesis in intact adult insects and diapausing silkworm eggs will be reviewed.

4.1. Antifreeze Production in Silkworm Eggs

Eggs of the silkworm *Bombyx mori* overwinter in a state of diapause. This condition is induced by the diapause hormone (DH) produced by the suboesophageal ganglion (SOG) of the mother. Eggs that are in diapause are characterized by a high concentration of sorbitol and glycerol (Chino, 1958), the function of which is the prevention of freezing of the tissues. The inverse relationship between the concentration of glycogen and that of sorbitol and glycerol during diapause development has been used to explain the origin of the polyols. Kageyama (1976), has also shown that there is a small but significant rise in the concentration of lactate and that the initial rate of sorbitol synthesis is much greater than that of glycerol. These findings have been confirmed by Yaginuma and Yamashita (1978), who followed the synthesis of sorbitol and glycerol from glycogen prelabelled with ^{14}C. When diapause development is completed, the polyols are reconverted into glycogen, the synthesis of which proceeds more rapidly from sorbitol than from glycerol. In addition to polyol accumulation, the diapause condition in the egg is characterized by a very low PFK activity (Kageyama and Ohnishi, 1971). The low activity of this enzyme undoubtedly influences the accumulation of sorbitol and glycerol, although the mechanism by which it might occur is obscure.

4.1.1. Polyol Precursors

Ovaries that produce diapause eggs have a much higher glycogen concentration than those producing nondiapause eggs (38.6 mg/g compared with 24.4 mg/g) (Chino, 1957; Yamashita and Hasagawa, 1964). The higher concentration of glycogen in ovaries destined to produce diapause eggs has been attributed to an effect of DH, since removal of the SOG from female pupae that would normally produce diapause eggs results in ovaries having 30% less glycogen (Hasegawa and Yamashita, 1965). Coincident with this effect, the level of trehalose in the blood rises from 4.61 mg/ml to 6.27 mg/ml. The concentration of glycogen in the fat body also rises significantly. The ovary exhibits a strong preference for trehalose; the rate of absorption being at least fivefold greater than for any other disaccharide tested (Shimada and Yamashita, 1979). The rise in hemolymph trehalose and reduced glycogen synthesis in the ovary when DH is absent suggests that the hormone acts to facilitate utilization of trehalose by the ovary rather than synthesis by the fat body. The direct action of DH on the ovary has been confirmed by implantation experiments utilizing male hosts.

4.1.2. Control of Ovarian Trehalase Activity

When the maternal SOG is removed, the decline in ovarian glycogen is accompanied by a corresponding decrease in trehalase activity of the tissue. The two events are temporally out of phase by one day (Yamashita *et al.*, 1972) such that the decrease in trehalase activity precedes the fall in glycogen concentration. The decrease in trehalase activity following removal of the SOG is due to the absence of DH. This is shown by the fact that injection of DH extract into female pupae that had previously had the SOG removed restores trehalase activity to the normal level within one day. This effect does not occur *in vitro*, at least in the short term. Thus, it appears likely that the effect of DH is to increase the rate of synthesis of new enzyme protein rather than activate preexisting enzyme. That trehalase is a principal site of control for glycogen synthesis by DH is also shown by the fact that SOG removal causes major impairment to the utilization of trehalose by the ovaries but not that of glucose (cited in Shimada and Yamashita, 1979). Since glucose is the only product resulting from the hydrolysis of trehalose, it is clear that trehalase is an important control site for glycogen synthesis. The synthesis of oocyte glycogen coincides with the appearance of the follicle cells that surround the oocytes. Shimada and Yamashita (1979) claim that ovarian trehalase activity is located at the surface membranes of the follicle cells. This suggests that trehalose derived from the hemolymph is hydrolyzed before entering the follicle cells; a view which is supported by the observation that trehalose does not accumulate within the ovarian cells (Shimada and Yamashita, 1979). The extracellular position of trehalase constitutes an important element in the regulation of glycogen synthesis in the cell because it could be used to control the rate at which glucose enters the cell. Any change in trehalase activity will therefore alter the glucose gradient between the intracellular and extracellular fluid compartments and consequently the rate of entry into the cell. Under these circumstances, the role of DH as a regulator of glycogen synthesis is easily understood.

4.1.3. Polyol Synthesis and Phosphofructokinase

The reason for the accumulation of polyols during diapause is still largely unknown. Early studies by Chino (1963) showed that during diapause there is a sharp reduction in the activity of the electron transport system. Under these circumstances, the glycolytic pathway is the principal source of ATP, and as a result, there is an accumulation of lactate.

The source of NAD^+, which makes possible the generation of ATP by the glycolytic pathway, is the reduction of dihydroxyacetone phosphate to α-glycerophosphate. The latter product is dephosphorylated by an acid phosphatase present in the egg (Chino, 1961), and the free glycerol that results accumulates in the egg.

Cessation of, or sharply reduced oxidative phosphorylation activity cannot be the sole reason for the accumulation of polyols, although it probably plays a significant role. When the absence of oxidative phosphorylation is mimicked by subjecting nondiapause eggs to anaerobic conditions, the concentration of lactate and polyols increases, as it does in diapause eggs when oxygen is available (Kageyama and Ohnishi, 1973). Anaerobic conditions, however, lead only to an accumulation of polyols during the first three days of egg development. It is this fact that precludes the possibility that a lesion in the electron transport system is the only factor involved in the stimulation of polyol synthesis. A second factor appearing to have an important influence on the synthesis of polyols during the first three days after oviposition is the apparent absence of PFK, both in diapause and nondiapause eggs (Kageyama and Ohnishi, 1971). Thus, the combination of a failure in electron transport and the absence of PFK activity may be the key to understanding polyol synthesis. The absence of PFk during the first three days of egg development in nondiapause eggs would also explain why anaerobiosis is effective in stimulating polyol synthesis only during this period.

The patterns of glycerol and sorbitol accumulation in diapause eggs are not identical (Yaginuma and Yamashita, 1978), which suggests that their synthesis is controlled by separate mechanisms. Although it has not been proven that an increase in anaerobic metabolism is the cause of the glycerol accumulation, the finding that anaerobic metabolism in flight muscle leads to an increase in α-glycerophosphate, the immediate precursor of glycerol, suggests that it is. The reason for the increase in sorbitol is not so easily explained. Polyol dehydrogenases, the enzymes responsible for the synthesis of sorbitol, are present in the egg (Chino, 1960). The reactions catalyzed by the enzymes are as follows:

$$\text{Glucose} + \text{NADPH} + H^+ \leftrightharpoons \text{Sorbitol} + \text{NADP}^+$$

$$\text{Glucose-6-PO}_4 \text{ (or Fructose-6-PO}_4\text{)} + \text{NADPH} + H^+ \rightleftharpoons \text{Sorbitol-6-PO}_4 + \text{NADP}^+$$

The sorbitol-6-PO_4 is dephosphorylated by the same acid phosphatase that dephosphorylates α-glycerophosphate (Chino, 1961). The polyol dehydrogenases are present in both types of eggs, thus the question is, why does sorbitol only accumulate in those eggs that are in the diapause state?

One possibility is the reduced PFK activity in those eggs (Kageyama and Ohnishi, 1973) having a low rate of electron transport. Reduced electron transport would be expected to stimulate glycolytic activity. This assumption appears to be correct, since lactate accumulates during the early stages of diapause (Kageyama, 1976). However, since PFK activity is markedly reduced (Kageyama and Ohnishi, 1973), glucose-6-phosphate that results from the phosphorolytic cleavage of glycogen will probably be redirected through the pentose phosphate pathway before being readmitted to the glycolytic pathway. This would make possible the synthesis of NADPH required for the polyol dehydrogenase reaction. The flow of reserve carbohydrate from glycogen to sorbitol and glycerol and their relationship to glycolysis and the pentose phosphate pathway is shown in Fig. 5. In summary, it can be stated as a tentative hypothesis that sorbitol synthesis occurs during egg diapause because of enhanced glycolytic activity resulting from impaired electron transport, but it also entails a reduction in PFK activity leading to greater oxidation of glucose-6-phosphate by the pentose phosphate pathway, thus providing the NADPH required for sorbitol synthesis. The very marked increase in sorbitol synthesis at the beginning of diapause suggests that a significant change in the equilibrium of the reactions favoring sorbitol synthesis from glycogen occurs at this time. The activation of phosphorylase at the be-

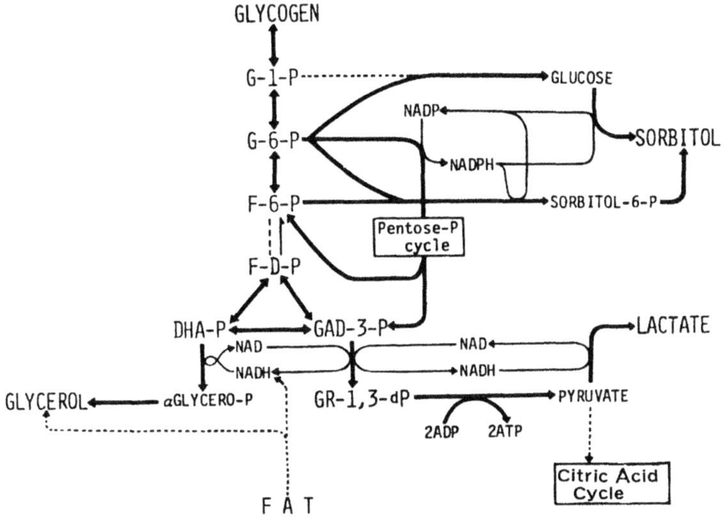

Fig. 5. Metabolic pathways for the synthesis of sorbitol, glycerol, and lactate in the diapausing silkmoth egg. The same pathways may also be utilized for polyol synthesis in the tissues of adult insects. Reproduced with permission from Kageyama and Ohnishi (1973) and the publisher.

ginning of diapause (Yamashita *et al.*, 1975) is undoubtedly responsible for this. Unfortunately, the reason for the activation is not known. The role of reduced (or absence of) PFK activity in the synthesis of sorbitol has recently been questioned, because studies in which glycogen in the diapause egg has been prelabeled with ^{14}C have shown that PFK activity is not absent, as previously thought, and glycolysis has been demonstrated to occur. Since only eggs in the diapause condition were examined, it is possible that diapause results in a reduction of PFK activity rather than a cessation. Because PFK is a rate-limiting enzyme in glycolysis, even small changes in activity might significantly affect the rate at which glucose-6-phosphate is channeled through the pentose phosphate pathway.

On completion of diapause, the accumulated sorbitol quickly disappears. This is accomplished by the appearance of NAD-dependent sorbitol dehydrogenase (Yaginuma and Yamashita, 1979). The synthesis or activation of this enzyme in diapause eggs appears to be absolutely dependent on exposure of the eggs to low temperature.

4.2. Polyol Synthesis in Larval and Adult Insects

Almost simultaneous with the discovery by Chino (1957) that diapausing silkworm eggs accumulated glycerol, Wyatt and Meyer (1959) demonstrated the accumulation of glycerol in the diapausing pupa of the silkmoth *Hyalophora cecropia*. Salt (1959) described a similar situation in overwintering larvae of *Bracon cephi*. Since then, the accumulation of glycerol, and also of sorbitol, has been shown to occur in a variety of insects that enter diapause or exhibit a tolerance for low temperatures. These include carpenter ants (Sømme (1964), a carabid beetle (Baust and Miller, 1970), the woolly bear (Mansingh and Smallman, 1972), and a blowfly (Wood and Nordin, 1976).

4.2.1. Polyols as Supercooling Agents

Salt (1959) was among the first to direct attention to the possibility that glycerol was the effective agent in the cold-hardening of insects. He demonstrated that as glycerol accumulated in the hemolymph of *Bracon cephi* the melting point was depressed as low as −17.5°C and the supercooling point to −47.2°C. Coincident with these changes, the glycerol content of the hemolymph rose to levels as high as 5 molal. It is the marked effect of glycerol on the supercooling point that enables the insect to avoid freezing. The relationship that exists between supercooling and the accumulation of glycerol (or sorbitol) in the blood has been shown for various species (Sømme, 1964; Baust and Miller, 1970; Mansingh and Smallman, 1972). It is important to note, however, that although the presence of glycerol may prevent the occurrence of freezing, it does not necessarily protect the insect once freezing has taken place (Sømme, 1964). The relationship between supercooling and seasonal changes in glycerol content in the carabid beetle (Pterostichus brevicornis) is shown in Fig. 6.

4.2.2. The Stimulus for Polyol Synthesis

The induction of polyol synthesis is not achieved by the same mechanism in all insects where it occurs. In certain instances low temperature alone is sufficient to promote synthesis of polyols. The stimulatory effect of low temperature on glycerol production has been demonstrated in the adults of carpenter ants (Dubach *et al.*, 1959), beetles (Baust and Miller, 1970), and blowflies (Wood and Nordin, 1976). Alternatively, glycerol and sorbitol synthesis can be increased in certain species by the special conditions that define the state of diapause, but a high temperature. This has been shown for the diapausing pupa of the silkmoth (Wyatt and Meyer, 1959; Ziegler and Wyatt, 1975) and woolly bear (Mansingh and Smallman, 1972). In both species, however, the rate of polyol synthesis is greatly increased if the temperature is lowered coincident with the onset of diapause. In insects undergoing diapause there appears to be a synergistic relationship between the diapause condition and low temperature where polyol synthesis is concerned. This relationship can be seen in the woolly

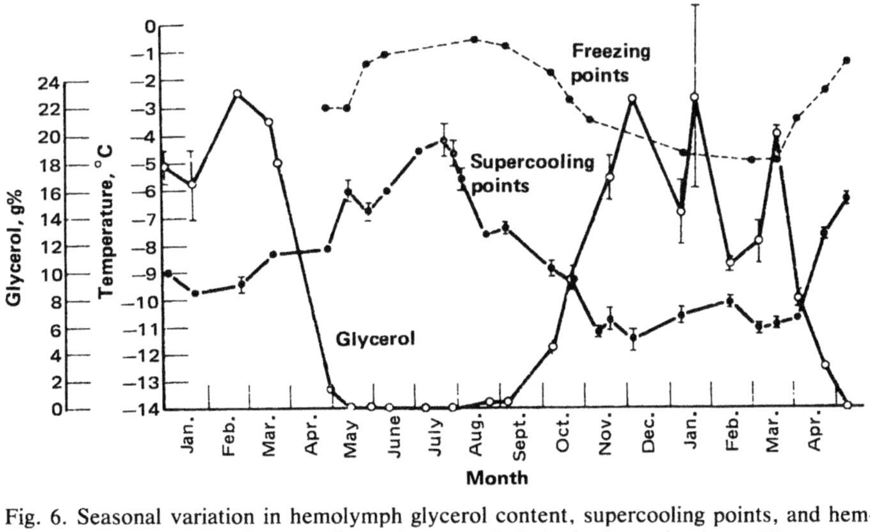

Fig. 6. Seasonal variation in hemolymph glycerol content, supercooling points, and hemolymph freezing points. Supercooling points are whole body determinations. All measurements are for adult *Pterostichus brevicornis*. Values are mean ± S.E. From Baust and Miller (1970). Reproduced with permission of the authors and Pergamon Press, Oxford.

bear, where completion of diapause reduces the rate of glycerol and sorbitol below that observed when low temperature and diapause are acting in concert.

4.2.3. The Mechanism for Polyol Synthesis

The reasons underlying the stimulation of glycerol and sorbitol synthesis in postembryonic insects are still vague, but an interesting beginning has been made in finding a solution to this intriguing problem. Polyol synthesis in larval and adult insects follows the same pathways as those already described for silkworm eggs (Fig. 5). The study by Ziegler *et al.* (1979) on the diapausing silkmoth pupa has shown that the major site of polyol synthesis, if not the only one, is the fat body.

Glycerol and sorbitol accumulating during diapause or periods of low temperatures are synthesized at the expense of glycogen (Mansingh and Smallman, 1972; Wood and Nordin, 1976). When diapause is completed or the period of exposure to low temperature terminated, the polyols are reconverted into glycogen (Wyatt, 1967; Mansingh and Smallman, 1972; Wood and Nordin, 1976). The fact that much of the glycogen giving rise to glycerol and sorbitol is later reconstituted from the polyols indicates that only a small part of the glycogen is completely oxidized. This suggests that glycogenolysis is not under respiratory control or subject to regulation by changes in the energy supply at this time. What then is the driving force that sends more glucosyl units through the glycolytic pathway in the absence of a demand for oxidizable substrate? The answer seems to be that fat body phosphorylase activity is much increased by low temperature. Ziegler and Wyatt (1975) and Ziegler *et al.* (1979) have shown unequivocally that lowering the temperature of silkmoth pupal fat body to O°C has a strong activating effect on the conversion of phosphorylase b to a (Fig. 2). This effect can be reversed by raising the temperature. Furthermore, the activating effect of low temperature has been demonstrated in the intact diapausing pupa, where it has been shown to coincide with the increase in glycerol levels. Although the activation of phosphorylase can explain the conversion of glycogen into glycerol, the means by which low temperature activates phosphorylase is unknown. Ashida and Wyatt (1979) have shown that pupal fat body phosphorylase b is converted to the a form by phosphorylase kinase but have been unable to show activation by low temperature. The activation of phosphorylase by low temperature does not require the special conditions associated with diapause. This is clearly shown by the fact that fat body phosphorylase activity increases at least 150% when locusts and insects are chilled at 4°C for 2 hr (Ziegler *et al.*, 1979). in our studies we have shown also that cockroach fat body responds to low temperature in a similar manner

(Dean and Steele, unpublished observations). These results suggest that low temperature activation of phosphorylase may be a more common occurrence than hitherto suspected. The phenomenon is of considerable biochemical interest and deserving of further study.

Wood and Nordin (1980) have described an interesting metabolic effect produced by low temperature in *Protophormia terranovae*. This blowfly produces glycerol but not sorbitol in response to cold stress, whereas *Musca domestica*, with which they compared the blowfly, produces neither. When the temperature is decreased to 2°C, the rate of glycolysis relative to that of the TCA cycle increases; the increment in the blowfly being slightly greater. The blowfly, however, exhibited a large increase in pentose pathway activity relative to that of the TCA cycle (Wood and Nordin, 1980). This effect did not occur in the housefly. The reason for the relative increase in pentose pathway activity in the cold-stressed insect is an enigma, since this particular species produces glycerol rather than sorbitol. There is therefore no obvious need for the relatively greater amount of NADPH that would be produced. In theory, at least, NADPH could transfer its reducing equivalents to NAD by means of a transhydrogenase so that the NADH so formed could participate in the glycerol-3-phosphate dehydrogenase reaction. This reaction is of potential importance as far as the synthesis of glycerol is concerned because of the presence in *Protophormia* of a phosphatase capable of hydrolysing glycerol-3-phosphate. The possibility that glycerol-3-phosphate is the source of the glycerol is suggested by the fact that the phosphatase shows enhanced activity in cold-stressed flies (Wood *et al.*, 1977).

5. Energy Metabolism in the Rectum

Desiccation is a potential hazard to many terrestrial insects. For this reason, the absorption of water from the feces to replace that lost by evaporation from the body surface is an important aspect of survival. It has been shown that transport of water across the wall of the locust rectum over short periods does not require ionized solute in the lumen fluid (Phillips, 1964). Ion transport is implied to occur, however, because the transepithelial water movement is blocked by various respiratory inhibitors (Vietinghoff, 1965; Wall, 1967; Irvine and Phillips, 1971). Since many insects living under dry conditions are transporting water out of the rectum against a concentration gradient, it is obvious that the mechanism is energy dependent.

Numerous studies have shown that water uptake from the rectum is considerably stimulated by antidiuretic factors (ADH) from the nervous system. Such factors have been described in the pars intercerebralis, corpora cardiaca, corpora allata, prothoracic ganglion, and terminal abdominal ganglion (Vietinghoff, 1967; Wall, 1967; Cazal and Girardie, 1968; de Bessé and Cazal, 1968; Mordue, 1970). Because these factors provide a marked enhancement of water transport, and in some instances even reverse lumen-directed flow of water, they can be extremely useful to the investigator who wishes to study the relationship between water movement and metabolism.

5.1. Na^+ Requirement for ADH Stimulated Water Transport

Wall (1967) has shown that water transport across the rectal epithelium of the cockroach, *Periplaneta americana*, from lumen to hemocoel is stimulated by extracts of various parts of the nervous system. Among the most active were the corpora allata. In these experiments the rectal lumen contained no ionized solute, but the preparation was bathed in Ringer solution containing Na^+. Subsequent studies by Steele and Tolman (1980) showed that the stimulating effect of the gland extract is absolutely dependent on Na^+ in the Ringer solution bathing the rectal preparation. The replacement of Na^+ with choline, however, does not interfere with the basal level of water transport that occurs in the absence of ADH. The requirement for Na^+ suggests that a (Na^+–K^+)-activated ATPase might be part of the hormone stimulated mechanism. This notion is confirmed by the finding that ADH stimulated water transport is completely blocked by the addition of 2.0 mM ouabain to the Ringer solution bathing the rectum (Steele and Tolman, 1980). Wall (1967) found that dehydration lowered the content of antidiuretic factor in the corpora cardiaca and corpora allata. This means, presumably, that cockroaches so treated had released the antidiuretic factor into the hemolymph, where it then had access to the rectum. It was found that recta taken from dehydrated cockroaches, and incubated without ADH, would transport water at the same rate as those taken from hydrated individuals and incubated with ADH (Steele and Tolman, 1980). Furthermore, omission of Na^+ from the Ringer solution bathing the dehydrated recta reduced water transport to the same level as in recta from hydrated cockroaches not treated with ADH. The finding that the rectal tissue is particularly rich in (Na^+–K^+)-activated ATPase (Tolman and Steele, 1976) points strongly towards its involvement in ADH stimulated water transport.

5.2. Glycogen as an Energy Source for Water Transport

The high level of (Na^+–K^+)-activated ATPase in the rectum (Tolman and Steele, 1976), the Na^+ requirement for ADH stimulated water transport, and the recognition that water movement is energy dependent prompted an investigation into the source of the energy utilized. It seems likely that glycogen is the principal reserve used for this purpose (Fig. 7). These findings have been confirmed by *in vitro* studies, in which it has been shown that phosphorylase, the rate-limiting enzyme in glycogenolysis, is activated by gland extract containing the ADH (Steele and Tolman, 1980). The effect is blocked if Na^+ is omitted from the Ringer solution bathing the recta or 2.0 mM ouabain is added. These findings point clearly to a coupling, of undetermined nature, between the (Na^+–K^+)-activated ATPase and glycogenolysis.

5.3. Water Transport Associated Respiration

The depletion of glycogen reserves by treatment of the recta with gland extract containing ADH does not prove that it is utilized for the synthesis of ATP used to support water transport. Evidence for this function was obtained by measuring the effect of the ADH-containing extracts on the respiration of the isolated rectum. Tolman and Steele

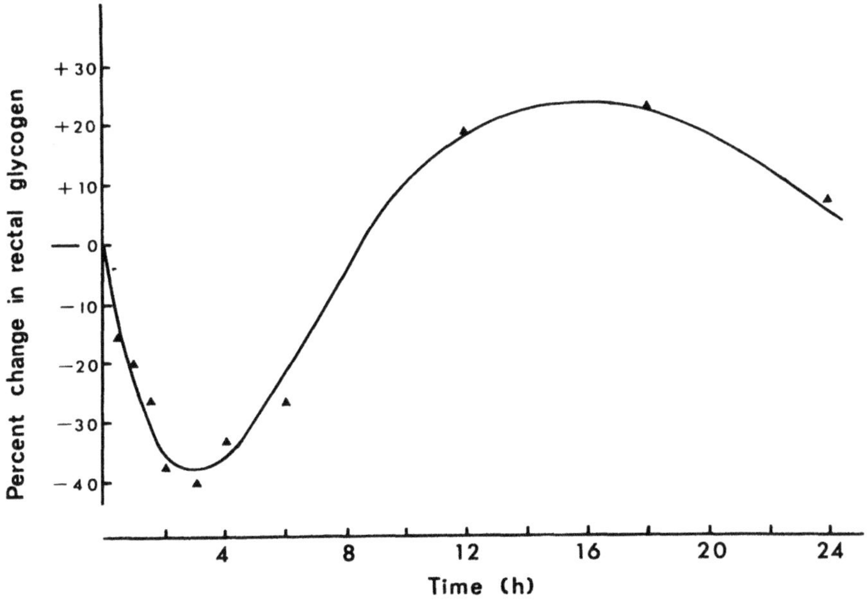

Fig. 7. The effect of corpora cardiaca–corpora allata extract on rectal glycogen levels *in vivo*. Cockroaches were injected with 10 μl cardiaca–allata extract, equivalent to 0.01 pair of glands in Ringer solution. The values for extract treated cockroaches are expressed as the difference (%) from the corresponding control group. The values for 1.5 to 6 hr are significantly different ($P < 0.05$). From Tolman and Steele (1980). Reproduced with permission of the authors and Pergamon Press, Oxford.

(1980) showed that extracts at concentrations up to 1 pair of glands per ml increase respiration by as much as 50%. Furthermore, this effect is concentration dependent. The stimulation of respiration evoked by the extract is completely abolished if Na^+ is omitted from the Ringer solution bathing the basal surface of the rectum (Tolman and Steele, 1980). Thus, these results complement and support the observations on glycogen depletion and phosphorylase activity.

The study shows that there is a close relationship between water transport and metabolism. Because the metabolic activity is considerable and very sensitive to hormonal regulation it may be an excellent system for the investigation of the coupling mechanism between physiological work and metabolism.

ACKNOWLEDGMENTS. Receipt of a research grant from the Natural Sciences and Engineering Research Council of Canada (NSERC) used to support certain of the studies cited in this review is gratefully acknowledged. I am also indebted to Mrs. Helen Kyle for transforming an illegible manuscript into readable typescript.

References

Ashida, M., and Wyatt, G.R., 1979, Properties and activation of phosphorylase kinase from silkmoth fat body, *Insect Biochem.* **9**:403.

Baust, J.G., and Miller, L.K., 1970, Variations in glycerol content and its influence on cold hardiness in the Alaskan carabid beetle, *Pterostichus brevicornis*, *J. Insect Physiol.* **16**:979.

Candy, D.J., and Kilby, B.A., 1961, The biosynthesis of trehalose in the locust fat body, *Biochem. J.* **78**:531.

Cazal, M., 1971, Action des corpora cardiaca sur la tréhalose et la glycémie de *Locusta migratoria* L, *C.R. Acad. Sci., Ser. D* **272**:2596.

Cazal, M., and Girardie, A., 1968, Control humoral de l'équilibre hydrique chez *Locusta migratoria migratoriodes*, *J. Insect Physiol.* **14**:655.

Chalaye, D., 1969, La tréhalosémie et son contrôle neuroendocrine chez le criquet migrateur *Locusta migratoria migratoriodes*. 2. Role de corpora cardiaca et des organes perisympathiques, *C.R. Acad. Sci., Ser. D* **268**:3111.

Chefurka, W., 1958, On the importance of α-glycerophosphate dehydrogenase in glycolysing insect muscle, *Biochem. Biophys. Acta* **28**:660.

Childress, C.C., and Sacktor, B., 1970, Regulation of glycogen metabolism in flight muscle. I. Purification and properties of phosphorylases. II. Kinetic properties and control of phosphorylase *in vivo*, *J. Biol. Chem.* **245**:2927.

Chino, H., 1957, Carbohydrate metabolism in diapause egg of the silkworm, *Bombyx mori*. I. Diapause and the change of glycogen content, *Embryologia* **3**:295.

Chino, H., 1958, Carbohydrate metabolism in diapause egg of the silkworm, *Bombyx mori*. II. Conversion of glycogen into sorbitol and glycerol during diapause, *J. Insect Physiol.* **2**:1.

Chino, H., 1960, Enzymatic pathways in the formation of sorbitol and glycerol in the diapausing egg of the silkworm *Bombyx mori*. I. On the polyol dehydrogenases, *J. Insect Physiol.* **5**:1.
Chino, H., 1961, Enzymatic pathways in the formation of sorbitol and glycerol in the diapausing egg of the silkworm, *Bombyx mori*. II. On the phosphatases, *J. Insect Physiol.* **6**:231.
Chino, H., 1963, Respiratory enzyme system of the *Bombyx* silkworm egg in relation to the mechanism of the formation of sugar alcohols, *Arch. Biochem. Biophys.* **102**:400.
Clegg, J.S., and Evans, D.R., 1961, Blood trehalose and flight metabolism in the blowfly, *Science* **134**:54.
Clements, A.N., Page, J., Borck, K., and van Ooyen, A.J.J., 1970, Trehalases of the fleshfly *Sarcophaga barbata, J. Insect Physiol.* **16**:1389.
Crabtree, B., and Newsholme, E.A., 1972, The activities of phosphorylase, hexokinase, phosphofructokinase, lactate dehydrogenase and the glycerol 3-phosphate dehydrogenases in muscles from vertebrates and invertebrates, *Biochem. J.* **126**:49.
Crabtree, B., and Newsholme, E.A., 1975, Comparative aspects of fuel utilization and metabolism by muscle, in: *Insect Muscle* (P.N.R. Usherwood, ed.), pp. 405–500, Academic Press, New York.
Davis, R.A., and Fraenkel, G., 1940, The oxygen consumption of flies during flight, *J. Exp. Biol.* **17**:402.
de Bessé, N., and Cazal, M., 1968, Action des extraits d'organes périsympathiques et de corpora cardiaca sur la diurese de quelques insectes, *C.R. Acad. Sci., Ser. D* **266**:615.
Downer, R.G.H., and Parker, G.H., 1979, Glycogen utilisation during flight in the American cockroach, *Periplaneta americana* L., *Comp. Biochem. Physio.* **64A**:29.
Dubach, P., Smith, F., Pratt, D., and Stewart, C.M., 1959, Possible role of glycerol in the winter-hardiness of insects, *Nature* (*London*) **184**:288.
Filburn, C.R., and Wyatt, G.R., 1976, Adenylate and guanylate cyclases of cecropia silkmoth fat body, *J. Insect Physiol.* **22**:1635.
Friedman, S., 1967, The control of trehalose synthesis in the blowfly, *Phormia regina* Meig., *J. Insect Physiol.* **13**:397.
Friedman, S., and Alexander, S., 1971, Multiple forms of trehalose in *Phormia regina*, *Biochem. Biophys. Res. Commun.* **42**:818.
Gilby, A.R., Wyatt, S.S., and Wyatt, G.R., 1967, Trehalases from the cockroach, *Blaberus discoidalis*: Activation, solubilization and properties of the muscle enzyme and some properties of the intestinal enzyme, *Acta Biochim. Pol.* **14**:83.
Goldsworthy, G.J., 1969, Hyperglycaemic factors from the corpus cardiacum of *Locusta migratoria, J. Insect Physiol.* **15**:2131.
Goldsworthy, G.J., 1970, The action of hyperglycaemic factors from the corpus cardiacum of *Locusta migratoria* on glycogen phosphorylase, *Gen. Comp. Endocrinol.* **14**:78.
Grasso, A., and Migliore-Natalizi, G., 1968, Studies on insect (*Periplaneta americana*) phosphofructokinase, *Comp. Biochem. Physiol.* **26**:979.
Gussin, A.E.S., and Wyatt, G.R., 1965, Membrane bound trehalose from *Cecropia* silkmoth muscle, *Arch. Biochem. Biophys.* **112**:626.
Hanaoka, K., and Takahashi, S.Y., 1976, Effect of a hyperglycaemic factor on haemolymph trehalose and fat body carbohydrates in the American cockroach, *Insect Biochem.* **6**:621.
Hanaoka K., and Takahashi, S.Y., 1977, Adenylate cyclase system and the hyperglycaemic factor in the cockroach, *Periplaneta americana*, *Insect Biochem.* **7**:95.
Hansen, K., 1966, Zur cytologischen lokalisation der trehalase in der indirekten flugmuskulatur der insekten, *Biochem. Z.* **344**:15.
Hansford, R.G., and Sacktor, B., 1970, Regulation of glycogen metabolism in insect flight

muscle. Activation of phosphorylase *b* kinase by calcium and inorganic phosphate, *FEBS Lett.* **7**:183.

Hasegawa, K., and Yamashita, O., 1965, Studies on the mode of action of the diapause hormone in the silkworm, *Bombyx mori* L. VI. The target organ of the diapause hormone, *J. Exp. Biol.* **43**:271.

Houben, N.M.D., and Beenakkers, A.M. T. 1975, The influence of haemolymph carbohydrate concentration on the release of adipokinetic hormone during locust flight, *J. Endocrinol.* **64**:66P.

Irvine, H.B., and Phillips, J.E., 1971, Effects of respiratory inhibitors and ouabain on water transport by isolated locust rectum, *J. Insect Physiol.* **17**:381.

Jones, J., Stone, J.V., and Mordue, W., 1977, The hyperglycaemic activity of locust adipokinetic hormone, *Physiol. Entomol.* **2**:185.

Jungreis, A.M., and Wyatt, G.R., 1972, Sugar release and penetration in insect fat body: Relations to regulation of haemolymph trehalose in developing stages of *Hyalophora cecropia, Biol. Bull. (Woods Hole, Mass.)* **143**:367.

Jutsum, A.R., and Goldsworthy, G.J., 1976, Fuels for flight in *Locusta, J. Insect Physiol.* **22**:243.

Kageyama, T., 1976, Pathways of carbohydrate metabolism in the eggs of the silkworm, *Bombyz mori, Insect Biochem.* **6**:507.

Kageyama, T., and Ohnishi, E., 1971, Carbohydrate metabolism in the eggs of the silkworm *Bombyx mori.* I. Absence of phosphofructokinase in the diapausing egg, *Dev. Growth Differ.* **13**:97.

Kageyama, T., and Ohnishi, E., 1973, Carbohydrate metabolism in the eggs of the silkworm, *Bombyx mori.* II. Anaerobiosis and polyol formation, *Dev. Growth Differ* **15**:47.

Krogh, A., and Weis-Fogh, T., 1951, The respiratory exchange of the desert locust (*Schistocerca gregaria*) before, during and after flight, *J. Exp. Biol.* **28**:344.

Mansingh, A., and Smallman, B.N., 1972, Variation in polyhydric alcohol in relation to diapause and cold-hardiness in the larvae of *Isia isabella, J. Insect Physiol.* **18**:1565.

Mordue, W., 1970, Evidence for the existence of diuretic and antidiuretic hormones in locusts, *J. Endocrinol.* **46**:119.

Mordue, W., and Goldsworthy, G.J., 1969, The physiological effects of corpus cardiacum extracts in locusts, *Gen. Comp. Endocrinol.* **12**:360.

Morishima, I., 1978, Adenylate cyclase in silkworm. Properties of the enzyme in pupal fat body, *J. Biochem.* (*Tokyo*) **84**:1495.

Morishima, I., 1979, Adenylate cyclase in silkworm: Regulation by calcium and calcium-binding protein, *Agric. Biol. Chem.* **43**:1127.

Murphy, T.A., and Wyatt, G.R., 1965, The enzymes of glycogen and trehalose synthesis in silkmoth fat body, *J. Biol. Chem.* **240**:1500.

Newsholme, E.A., and Start, C., 1973, *Regulation in Metabolism*, John Wiley and Sons, London.

Newsholme, E.A., and Taylor, K., 1969, Glycerol kinase activities in muscles from vertebrates and invertebrates, *Biochem. J.* **112**:465.

Phillips, J.E., 1964, Rectal absorption in the desert *Locust, Schistocerca gregaria* Forskål. I. Water, *J. Exp. Biol.* **41**:15.

Polacek, L., and Kubista, V., 1960, Metabolism of the cockroach, *Periplaneta americana* during flight, *Physiol. Bohemoslov.* **9**:228.

Reed, W.D., and Sacktor, B., 1971, Localization of trehalase in flight muscle of the blowfly *Phormia regina, Arch. Biochem. Biophys.* **145**:392.

Rowan, A.N., and Newsholme, E.A., 1979, Changes in the contents of adenine nucleotides and intermediates of glycolysis and the citric acid cycle in flight muscle of the locust upon flight and their relationship to the control of the cycle, *Biochem. J.* **178**:209.

Sacktor, B., 1965, Energetics and respiratory metabolism of muscular contraction, in: *The Physiology of Insecta* (M. Rockstein, ed.), Volume 2, pp. 483–580, Academic Press, New York.

Sacktor, B., 1970, regulation of intermediary metabolism, with special reference to the control mechanisms in insect flight muscle, in: *Advances in Insect Physiology*, Vol. 7 (J.W.L. Beament, J.E. Treherne, and V.B. Wigglesworth, eds.), pp. 267–347, Academic Press, London.

Sacktor, B., 1975, Biochemistry of insect flight, in: *Insect Biochemistry and Function* (D.J. Candy, and B.A. Kilby, eds.), pp. 1–88, Chapman and Hall, London.

Sacktor, B., and Hurlbut, E.C., 1966, Regulation of metabolism in working muscle *in vivo*. II. Concentrations of adenine nucleotides, arginine phosphate and inorganic phosphate in insect flight muscle during flight, *J. Biol. Chem.* **241**:632.

Sacktor, B., and Wormser-Shavit, E., 1966, Regulation of metabolism in working muscle *in vivo*. I. Concentration of some glycolytic, tricarboxylic acid cycle, and amino acid intermediates in insect flight muscle during flight, *J. Biol. Chem.* **241**:624.

Sacktor, B., Wu, N.-C., Lescure, D., and Reed, W.D., 1974, Regulation of muscle phosphorylase *b* kinase activity by inorganic phosphate and calcium ions, *Biochem. J.* **137**:535.

Salt, R.W., 1959, Role of glycerol in the cold-hardening of *Bracon cephi* (Gahan), *Can. J. Zool.* **37**:59.

Shimada, S., and Yamashita, O., 1979, Trehalose absorption related with trehalase in developing ovaries of the silkworm, *Bombyx mori*, *J. Comp. Physiol.* **131**:133.

Sømme, L., 1964, Effects of glycerol on cold-hardiness in insects, *Can. J. Zool.*, **42**:87.

Steele, J.E., 1963, The site of action of insect hyperglycaemic hormone, *Gen. Comp. Endocrinol.* **3**:46.

Steele, J.E., 1964, The activation of phosphorylase in an insect by adenosine 3′,5′-phosphate and other agents, *Am. Zool.* **4**:328.

Steele, J.E., 1980, Hormonal modulation of carbohydrate and lipid metabolism in fat body, in: *Insect Biology in the Future* (M. Locke and D.S. Smith, eds.), pp. 253–271, Academic Press, New York.

Steele, J.E., and Tolman, J.H., 1980, Regulation of water transport in the cockroach rectum by the corpora cardiaca–corpora allata system: The requirement for Na^+, *J. Comp. Physiol.* **13**:357.

Stevenson, E., 1968, Carbohydrate metabolism in the flight muscle of the southern army worm moth, *Prodenia eridania*, *J. Insect Physiol.* **14**:179.

Stevenson, E., and Wyatt, G.R., 1964, Glycogen phosphorylase and its activation in silkmoth fat body, *Arch. Biochem. Biophys.* **108**:420.

Sugden, P.H., and Newsholme, E.A., 1975, The effects of ammonium, inorganic phosphate and potassium ions on the activity of phosphofructokinases from muscle and nervous tissues of vertebrates and invertebrates, *Biochem. J.* **150**:113.

Takahashi, S.Y., and Hanaoka, K., 1977, Multiple protein kinases in the American cockroach *Periplaneta americana*, *Insect Biochem.* **7**:133.

Tolman, J.H., and Steele, J.E., 1976, A ouabain sensitive (Na^+–K^+)-activated ATPase in the rectal epithelium of the American cockroach, *Periplaneta americana*, *Insect Biochem.* **6**:513.

Tolman, J.H., and Steele, J.E., 1980, The effect of the corpora cardiaca–corpora allata system on oxygen consumption in the cockroach rectum. The role of Na^+ and K^+, *J. Comp. Physiol.* **66B**:59.

Treherne, J.E., 1958, The absorption and metabolism of some sugars in the locust, *Schistocerca gregaria* (Forsk.), *J. Exp. Biol.* **35**:611.

Van Handel, E., and Nayar, J.K., 1973, Direct use of carbohydrates during sustained flight in the moth, *Spodoptera frugiperda, Insect Biochem.* **2:**203.

Vaughan, H., Thornton, S.D., and Newsholme, E.A., 1973, The effects of calcium ions on the activites of trehalase, hexokinase, phosphofructokinase, fructose diphosphatase and pyruvate kinase from various muscles, *Biochem. J.* **132:**527.

Vietinghoff, V., 1965, Untersuchungen über die Funktion der Rectaldrüsen der Stabheuschrecke *Carausius morosus* Br., *Zool. Anz.* (Suppl.) **29:**157.

Vietinghoff, V., 1967, Neurohormonal control of "renal function" in *Carausius morosus* Br., *Gen. Comp. Endocrinol.* **9:**503.

Walker, P.R., and Bailey, E., 1969, A comparison of the properties of the phosphofructokinases of the fat body and flight muscle of the adult male desert locust, *Biochem. J.* **111:**365.

Wall, B.J., 1967, Evidence for antidiuretic control of rectal water absorption in the cockroach, *Periplaneta americana* L. *J. Insect. Physiol.* **13:**565.

Weis-Fogh, T., 1964, Diffusion in insect wing muscle, the most active tissue known, *J. Exp. Biol.* **41:**229.

Wiens, A.W., and Gilbert, L.I., 1965, Regulation of cockroach fat-body metabolism by the corpus cardiacum *in vitro, Science* **150:**614.

Wiens, A.W., and Gilbert, L.I., 1967, The phosphorylase system of the silkmoth, *Hyalophora cecropia, Comp. Biochem. Physiol.* **21:**145.

Wood, F.E., and Nordin, J.H., 1976, Studies on the low temperature induced biogenesis of glycerol by adult *Protophormia terranovae, J. Insect Physiol.* **22:**1665.

Wood, F.E., and Nordin, J.H., 1980, Activation of hexose monophosphate shunt during cold-induced glycerol accumulation by *Protophormia terranovae, Insect Biochem.* **10:**87.

Wood, F.E., Mahar, P., Nordin, J.H., 1977, Metabolite levels and enzyme activities in *Protophormia terranovae* during low temperature induced glycerol accumulation, *Insect Biochem.* **7:**141.

Worm, R.A.A., and Beenakkers, A.M. T., 1980, Regulation of substrate utilization in the flight muscle of the locust, *Locusta migratoria* during flight, *Insect Biochem.* **10:**53.

Wyatt, G.R., 1967, The biochemistry of sugars and polysaccharides in insects, in: *Advances in Insect Physiology* (J.W.L. Beament, J.E. Treherne, and V.B. Wigglesworth, eds.), Volume 4, pp. 287–360, Academic Press, London.

Wyatt, G.R., and Meyer, W.L., 1959, The chemistry of insect haemolymph. III. Glycerol, *J. Gen. Physiol.* **42:**1005.

Yaginuma, T., and Yamashita, O., 1978, Polyol metabolism related to diapause in *Bombyx* eggs: Different behavior of sorbitol from glycerol during diapause and post-diapause, *J. Insect Physiol.* **24:**347.

Yaginuma, T., and Yamashita, O., 1979, NAD-dependent sorbitol dehydrogenase activity in relation to the termination of diapause in eggs of *Bombyx mori, Insect Biochem.* **9:**547.

Yamashita, O., and Hasegawa, K., 1964. Studies on the mode of action of the diapause hormone in the silkworm, *Bombyx mori* L. IV. Effect of diapause hormone on the glycogen content in ovaries and the blood sugar level of silkworm pupae, *J. Sericult. Sci.* **33:**407.

Yamashita, O., Hasegawa, K., and Seki, M., 1972, Effect of the diapause hormone on trehalase activity in pupal ovaries of the silkworm *Bombyx mori* L., *Gen. Comp. Endocrinol.* **18:**515.

Yamashita, O., Suzuki, K., and Hasegawa, K., 1975, Glycogen phosphorylase activity in relation to diapause initiation in *Bombyx* eggs, *Insect Biochem.* **5:**707.

Yanagawa, H.-A., and Horie, Y., 1978, Activating enzyme of phosphorylase *b* in the fat body of the silkworm, *Bombyx mori, Insect Biochem.* **8**:155.
Zebe, E., 1954, Über den Stoffwechsel der Leipidopteren, *Z. Vgl. Physiol.* **36**:290.
Zebe, E.C., and McShan, W.H., 1957, Lactic and α-glycerophosphate dehydrogenases in insects, *J. Gen. Physiol.* **40**:779.
Zebe, E.C., and McShan, W.H., 1959, Incorporation of ^{14}C-acetate into long chain fatty acids by the fat body of *Prodenia eridania, Biochim. Biophys. Acta*, **31**:513.
Ziegler, R., and Wyatt, G.R., 1975, Phosphorylase and glycerol production activated by cold in diapausing silkmoth pupae, *Nature* (*London*) **254**:622.
Ziegler, R., Ashida, M., Fallon, A.M., Wimer, L.T., Wyatt, S.S., and Wyatt, G.R., 1979, Regulation of glycogen phosphorylase in fat body of *cecropia* silkmoth pupae, *J. Comp. Physiol.* **131**:221.

5

The Role of Proline in Energy Metabolism

Einar Bursell

1. Introduction

Amino acids have long been recognized to be prominent constituents of the hemolymph and tissue fluids of insects, accounting for more than 30% of the total osmotic activity in advanced orders (Sutcliffe, 1963). Details of the amino acid pattern have been elucidated for many species, and the results have been widely reviewed (Florkin, 1958; Wyatt, 1961; Chen, 1966; Schoffeniels and Gilles, 1970; Jeuniaux, 1971), to reveal a situation of bewildering, but seemingly controlled (Collett, 1976a), complexity. The only feature that stands out as a general, though by no means universal, characteristic of insect hemolymph is the predominance of proline and/or of glutamate and its amide glutamine (see also more recent investigations of Barrett, 1974; Nurmi and Birt, 1974; Bailey, 1975; Barrett and Friend, 1975; Collett, 1976b; de Kort and Kramer, 1976.) The early work of Winteringham with radioactively labeled substrates indicated a high rate of turnover in amino acid pools (Winteringham and Harrison, 1956; Winteringham, 1958) and suggested that hemolymph amino acids in general, and proline and glutamate in particular, could serve as important sources of substrate for the Krebs cycle (see also reviews by Sacktor, 1961, 1974). Support for this view was provided by later work, which showed a significant depletion of proline during activity in a number of

Einar Bursell • Department of Zoology, University of Rhodesia, Salisbury, Rhodesia. Present address: Tsetse Research Laboratory, Department of Veterinary Medicine, Langford, near Bristol, England.

different insects (Bursell, 1963; Corrigan and Kearns, 1963; Kirsten *et al.*, 1963; Ray, 1964; Sacktor and Wormser-Shavit, 1966; Barker and Lehner, 1972; Brouwers and de Kort, 1979). Since then, proline metabolism has been under intensive investigation, and its importance in energy metabolism has become firmly established.

2. Proline Metabolism

2.1. The Utilization of Proline during Flight

The pattern and rate of proline utilization during insect flight varies enormously from species to species. In the tsetse fly (*Glossina morsitans*) proline reserves are rapidly and continuously depleted until flight ceases after a few minues (Bursell, 1978); in the potato beetle (*Leptinotarsa decemlineata*) and the stable fly (*Stomoxys calcitrans*) there is a rapid fall in proline content during the initial phase of flight, but activity is then sustained without further change in proline level (Lewis, 1976; Weeda *et al.*, 1979); whereas in the locust (*Locusta migratoria*) and the blowfly (*Phormia regina*), an initial fall in proline content is followed by a slow decline during prolonged flight (Kirsten *et al.*, 1963; Sacktor and Wormser-Shavit, 1966; Mayer and Candy, 1969). The maximum rate of utilization also shows wide variation, from about 90 μmol/min/g fresh weight in the tsetse fly to about 3 μmol/min/g fresh weight in the blowfly. These differences suggest that proline may play different roles in the flight metabolism of different species, and detailed investigations have been undertaken to provide a basis of interpretation.

2.2. Pathways of Proline Catabolism

2.2.1. The Tsetse Fly

Early indications concerning the pathway of proline oxidation were provided by the observation that the initial depletion of proline during activity is associated with a roughly stoichiometric increase in alanine (Bursell, 1963; Ray, 1964; Sacktor and Wormser-Shavit, 1966; Hargrove, 1976; Weeda *et al.*, 1979), suggesting that transamination is an important feature of the pathway (Fig. 1). In the tsetse fly there is also a transient increase in glutamate, and it was suggested that proline is first oxidized to glutamate; this is followed by transfer of the amino group to pyruvate to form alanine, with the resulting oxoglutarate undergoing oxidation through Krebs cycle intermediates to oxaloacetate (now known to be malate, see below), decarboxylation of which replaces the pyruvate orig-

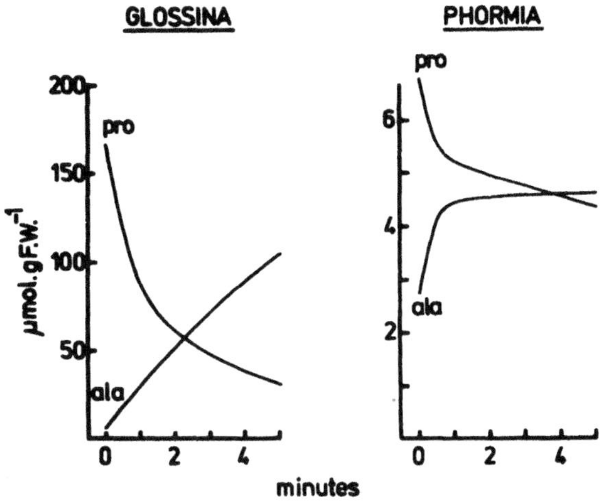

Fig. 1. Changes in proline and alanine content during flight in *Glossina pallidipes* (from Hargrove, 1976) and *Phormia regina* (from Sacktor and Wormser-Shavit, 1966). F.W.: fresh weight.

inally used as the amino acceptor. The pathway is illustrated in Fig. 2, and its operation was subsequently confirmed for the tsetse fly by the use of radioactive tracers (Bursell, 1966, 1967).

Later work has shown that during initial phases of flight there is a net loss of amino nitrogen (Hargrove, 1976), suggesting that a part of the glutamate formed from proline may be subject to oxidative deamination rather than to transamination. At this time, a portion of the pyruvate generated by the "malic" enzyme would not be required to serve as an amino acceptor in the aminotransferase reaction, and this pyruvate would therefore become available, through acetyl CoA for further oxidation in the Krebs cycle, as indicated by the dotted lines in Fig. 2.

The overall reaction for the partial oxidation of proline to alanine is

$$\text{proline} + 2\tfrac{1}{2}\ O_2 + 14\ \text{ADP} \rightarrow \text{alanine} + H_2O + 2\ CO_2 + 14\ \text{ATP}$$

and the yield of energy per unit mass of material oxidized is 0.52 mol ATP/g, compared with values of 0.18 for glucose and 0.65 for fat. Thus, in its partial combustion, proline combines the high energy yield of a lipid with the ready mobilizability of a carbohydrate, while the excessive build up of ammonia which would normally be associated with the rapid oxidation of amino acids is avoided by the use of a coupled transamination. For that portion of the proline that is totally oxidized, the energy yield per unit mass is reduced, but at 0.30 mol of ATP/g it is still well in excess of values for the total oxidation of carbohydrate.

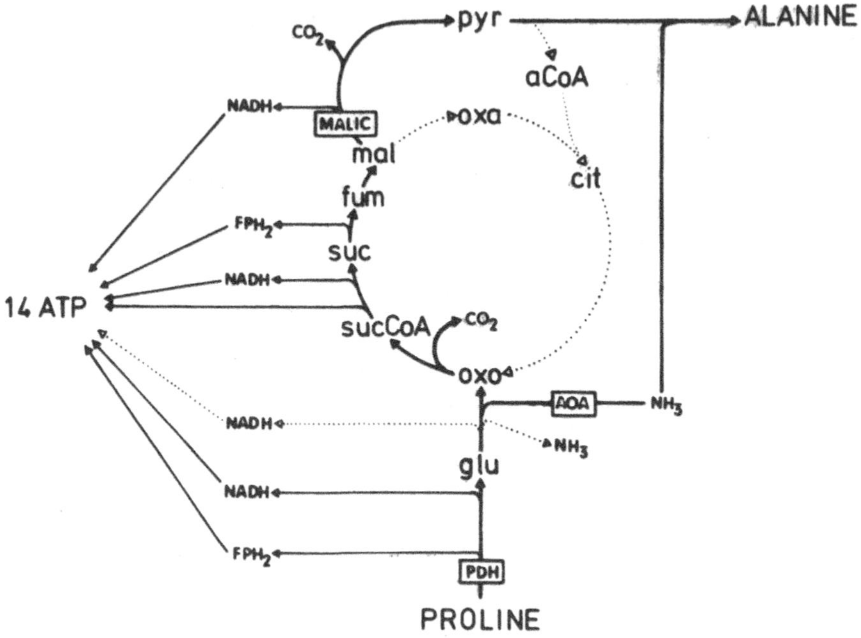

Fig. 2. Pathways of proline oxidation in the tsetse fly. Intermediate metabolites are denoted by the first three letters of their name, and enclosed letters show key enzymes of the system. Main pathways are indicated by heavy, and subsidiary pathways by dotted, lines. PDH: proline dehydrogenase; AOA: alanine-oxoglutarate aminotransferase; MALIC: "malic" enzyme; aCoA: acetyl CoA; FPH_2: reduced flavoprotein.

The properties of sarcosomes isolated from the flight muscles of the tsetse fly reflect the proline-centered metabolism of this insect in a number of ways. Only with proline can rates of oxidation be achieved that are commensurate with the requirements of flight metabolism; systems involved with the oxidation of carbohydrates and lipids are by comparison very poorly developed (Bursell and Slack, 1976), and it is clear that they can play little part in the provision of energy for flight. Alanine is a major product of the mitochondrial oxidation of proline, but it appears in less than stoichiometric proportion, which confirms the participation of glutamic dehydrogenase in the overall reaction. Under *in vitro* conditions, approximately 20% of the proline is subjected to complete oxidation following oxidative deamination, in approximate accord with the observed deficit in amino nitrogen during early phases of flight.

The pattern of enzyme activity in sarcosomes of the tsetse fly also conforms well to the postulated pathway. The interconversion of proline and glutamate is a two-step process involving an NAD-linked Δ-pyrroline-5-carboxylate dehydrogenase and a flavoprotein-linked proline dehydrogenase (Brosemer and Veerabhadrappa, 1965; Reddy and Campbell, 1969; Wadano *et al.*, 1976; Yoshida *et al.*, 1977), and the latter has a higher specific activity in the tsetse fly than in most other insects (Crabtree and Newsholme, 1970). The specific activities of alanine-oxoglutarate aminotransferase and of glutamate dehydrogenase are also high (Muus, 1968; Crabtree and Newsholme, 1970; Bursell, 1975a), as are those of the Krebs cycle enzymes involved with proline oxidation, as compared with their counterparts in the relatively inactive segment of the Krebs cycle (Norden and Matanganyidze, 1979). A primary requirement for the operation of the pathway is a mechanism for the rapid removal of substrate from the mitochondrion, which would otherwise be osmotically threatened by the input of metabolite, as noted by Hansford and Johnson (1975). It was originally thought that oxaloacetic decarboxylase was active in this respect (Bursell, 1965a), but it has since been shown that the enzyme responsible is an NAD-linked "malic" enzyme (Hoek *et al.*, 1976), which is highly active in sarcosomes of the tsetse fly, and so capable of competing with malate dehydrogenase to ensure removal of substrate from the Krebs cycle. The fact that this enzyme shows substantial activity in the decarboxylation of oxaloacetate (see also Lewis and Price, 1956; Norden and Matanganyidze, 1977) could perhaps be seen as a fail-safe device in the context of the osmotic stresses which might result from a failure to unload intramitochondrial substrate.

The importance of proline as a flight substrate is further underlined by the poor development of carbohydrate reserves (Geigy *et al.*, 1959; D'Costa *et al.*, 1973), by the low level of glycolytic metabolites (Birtwistle, 1974), by attenuation of the enzymatic machinery involved in carbohydrate metabolism (Norden and Paterson, 1969; 1970), and by indications that the direct oxidation of lipids is not a prominent feature of flight metabolism (Bursell *et al.*, 1974). Notwithstanding views to the contrary (Nayar and Van Handel, 1972), which can be shown to be based on mistaken assumptions concerning the energy requirement of flight metabolism, this leaves proline as the sole flight substrate, and the flight capacity of the insect is therefore effectively limited by the magnitude of its proline reserve (Brady, 1970; Bursell, 1978).

Although the general picture seems clear as far as the tsetse fly is concerned, there are a number of observations which have yet to be satisfactorily interpreted. Aspartic acid, for instance, does not feature anywhere in the metabolic pathway, yet in radioactive tracer experiments it is always strongly labeled (Bursell, 1963, 1966; Moloo *et al.*, 1974; Moloo, 1976a; 1977a,b). The apparent labeling of arginine, found in early experiments (Bursell, 1966), has not been confirmed by later work (Moloo, 1976a; 1977a,b); its original identification was based on paper chromatographic evidence and was clearly mistaken. The labeled component has since been separated from arginine by prolonged paper chromatographic development (Bursell, unpublished observation), but it has not yet been possible to establish its identity; the indications are that it is an unstable complex of a number of different amino acids, but the significance of its occurrence remains mysterious. Another puzzling feature is the similarity of results obtained when either glutamate or proline are used as labeled substrates for flight metabolism (Bursell, 1966), in view of later findings that sarcosomes of the tsetse fly have little capacity for glutamate oxidation (Bursell and Slack, 1976). Until these apparent discrepancies have been resolved, the concepts which have been developed concerning the pattern of tsetse metabolism must clearly remain provisional.

2.2.2. Comparative Aspects

Since proline appears to be of general importance as a constituent of insect hemolymph, the question arises whether in insects generally its oxidation conforms to the pattern established for the tsetse fly, and whether it fulfills a similar function in the supply of energy for flight. Although comprehensive information is seldom available, it should be possible to make some assessment of the situation by focusing attention on key aspects of the metabolic pattern, with particular reference to:

1. The relative capacities of sarcosomes for the oxidation of pyruvate and proline.
2. The specific activity of proline dehydrogenase, which would provide an indication of the input capacity.
3. The relative activity of the two competing enzymes, "malic" enzyme and malate dehydrogenase, which would reflect the sarcosomes output capacity.
4. The specific activity of alanine-oxoglutarate aminotransferase, which would provide an indication of the transamination potential.

Using these four criteria of a proline-dominated flight system it is possible to make a comparison between a number of different insects, and representative data are set out in Table 1. Values range widely for each of the four functions, with the tsetse fly at or near the upper extreme. Judging solely by the first criterion (Q_{O_2}Pro/Q_{O_2}Pyr) there is some indication that within the Diptera a capacity for rapid oxidation of proline may be associated with the bloodsucking mode of life (see Bursell, 1975b for further details), but this concept obviously has no general validity, since a capacity for proline oxidation is also well developed in phytophagous members of the Coleoptera, which show intermediate or high values for each of the four criteria.

Table 1. The Importance of Proline Oxidation in the Flight Metabolism of Different Species[a,b]

Order	Genus	Feeding habit	Q_{O_2}PRO/ Q_{O_2}PYR	PDH	ME/ MDH	AOA
Diptera	*Glossina*	♂ + ♀ Obligatory bloodsuckers	76.0	40	90	402
	Stomoxys	♂ + ♀ Facultative bloodsuckers	1.1	—	—	—
	Tabanus	♀ Facultative bloodsuckers	0.1	—	—	—
	Sarcophaga	Oligophagous	0.1	3	3	40
	Phormia	Oligophagous	0.2	—	—	—
Coleoptera	*Leptinotarsa*	Phytophagous	3.2	—	12	—
	Melolontha	Phytophagous	—	60	—	113
	Heliocarpus	Phytophagous	—	—	12	—
	Popillio	Phytophagous	1.2	100	—	1800
Dictyoptera	*Locusta*	Phytophagous	0.1	2	—	34
Orthoptera	*Periplaneta*	Oligophagous	—	7	1	45
Lepidoptera	*Laothoe*	Phytophagous	—	2	—	8
	Spodoptera	Phytophagous	—	—	1	—
	Prodenia	Phytophagous	0.1	—	—	—
Hymenoptera	*Apis*	Phytophagous	—	2	1	7

[a] Data from: Childress and Sacktor, 1966; Stevenson, 1968; Crabtree and Newsholme, 1970; de Kort *et al.*, 1973; Hansford and Johnson, 1975; Bursell, 1975b; Hoek *et al.*, 1976; Weeda *et al.*, 1980a.
[b] Judged by the relative capacity of sarcosomes for proline and pyruvate oxidation (Q_{O_2} PRO/Q_{O_2} PYR), the specific activity of proline dehydrogenase (PDH, μmol/min/g wet weight) the activity of "malic" enzyme as a percentage of the activity of malate dehydrogenase (ME/MDH), and the specific activity of alanine-oxoglutarate aminotransferase (AOA, μmol/min/g wet weight).

Details of the oxidation of proline by sarcosomes of the Japanese beetle, *Popillio japonica*, have been elucidated by Hansford and Johnson (1975), who show that the situation resembles that described for the tsetse fly except that the proportion of alanine recovered is much lower, indicating that despite the presence of a powerful aminotransferase, a much higher proportion of the proline is subjected to complete, rather than to partial, oxidation. A highly active "malic" enzyme ensures that the input of substrate to the mitochondrion is balanced by rapid decarboxylation of malate, thereby avoiding osmotic stress.

In the Colorado potato beetle, *Leptinotarsa decemlineata*, where the work of de Kort and his colleagues (de Kort *et al.*, 1973; Khan and de Kort, 1978; Mordue and de Kort, 1978; Brouwers and de Kort, 1979; Weeda *et al.*, 1979) indicates that proline serves as a primary substrate for flight, the oxidation of proline by isolated mitochondria results in the stoichiometric appearance of alanine (Weeda *et al.*, 1980a). The production of pyruvate at a rate commensurate with the transamination requirement is assured by the presence of a powerful "malic" enzyme, whereas glutamic dehydrogenase, which is known to be active in flight muscles of this species (Khan and de Kort, 1978), appears to play no part in the *in vitro* reaction.

All other species for which data are available resemble the nonblood-sucking Diptera, with extremely low input and output functions and relatively poor capacities for proline oxidation and transamination. In these, proline could only contribute marginally to the provision of energy for flight, yet in several of them a significant decrease in proline content has been demonstrated in the flying insect (Kirsten *et al.*, 1963; Sacktor and Wormser-Shavit, 1966; Mayer and Candy, 1969; Barker and Lehner, 1972; Worm and Beenakkers, 1980). Consideration has accordingly been given to the possibility that in these species proline might fulfil some other metabolic function. In his early work with sarcosomes of the housefly, van den Bergh (1964) noted that glutamate, and by implication proline, could serve as a source of Krebs cycle intermediates, although he noted that to the extent that a transamination with pyruvate was involved, it would do so at the expense of oxidizable substrate. This concept appeared to receive support from later work, which showed that the performance of mitochondria whose capacity to oxidize pyruvate had declined during storage could be restored by proline (Sacktor *et al.*, 1965; Sacktor and Childress, 1967; Stevenson, 1968; Slack and Bursell, 1976); it also provided a basis for interpretation of the transient increase in pyruvate concentration observed during the transition from rest to flight, which indicated that activation of oxidative mechanisms lagged behind that of the glycolytic pathway (Sacktor and Wormser-Shavit, 1966). All these effects could be explained in terms of an intramitochondrial generation of oxaloacetate, which would compensate for losses of Krebs substrate sustained during storage (or rest) and thereby increase the capacity of the system for condensation with acetyl CoA and hence for pyruvate oxidation. According to this view, the pyruvate used as an amino acceptor is thought to derive from glycolytic sources, but whether this is so has not been experimentally established. In experiments with isolated sarcosomes, alanine features quite prominently as a product of proline oxidation (Sacktor and Childress, 1967), indicating that, in the absence of glycolysis, pyruvate generated intramitochondrially is used as a substrate for transamination.

It has also been suggested that, in addition to its function as a source of Krebs cycle substrate, proline may play an important function as a reserve of glutamate for transamination with pyruvate, pending activation of oxidative pathways (Crabtree and Newsholme, 1975). In this way the large increase in pyruvate that would result from activation of the glycolytic pathway would be buffered by a rapid conversion of pyruvate to alanine, thereby stabilizing pyruvate concentration during the transition from rest to flight.

The idea that the primary function of proline oxidation in the blowfly is to provide an input of Krebs cycle substrate to substrate-depleted mitochondria appears now to have gained general acceptance (e.g., Sacktor, 1970; Hochachka and Somero, 1973; Crabtree and Newsholme, 1975; Kammer and Heinrich, 1978; Worm and Beenakkers, 1980), though the evidence on which it is based remains circumstantial. The restorative action of proline on pyruvate oxidation in aged mitochondria is admittedly a telling effect, but restored values are equal to, rather than in excess of, the values recorded with freshly isolated mitochondria (Sacktor and Childress, 1967), and for this reason the results provide no evidence that substrate depletion is of importance in the resting fly; indeed it is difficult to see how it could be, since the sarcoplasm is known to contain relatively high concentrations of Krebs cycle precursors such as proline, glutamate, and aspartate. Furthermore, the postulated function should be reflected in a sustained increase in the mitochondrial concentration of Krebs cycle substrate during activity; but neither oxaloacetate, citrate, nor oxoglutarate show any change during the transition from rest to flight, while the slight increase in malate concentration (from 0.6 to 0.9 μmol/g wet weight) is not commensurate with, nor does it follow the same time course as, the corresponding input of glutamate and proline, which totals 1.7 μmol/g wet weight during the first minute of flight. A similar discrepancy has been reported for the locust (Worm & Beenakkers, 1980), and although the apparent anomaly might be resolved on the basis of changes in the concentration of substrates that were not investigated (such as succinate or fumarate), evidence is lacking.

A completely different suggestion concerning the role of proline in blowfly flight metabolism has been made by Balboni (1978), based on the existence of an active NAD-linked Δ-pyrroline-5-carboxylate reductase in the sarcoplasm of the flight musculature. The proline produced by this reaction would be subject to reoxidation by mitochondrial proline dehydrogenase, thereby reconstituting Δ-pyrroline-5-carboxylate for return to the sarcoplasm. The cyclic system would thus serve as a shuttle mechanism for the oxidation of extramitochondrial NADH, analogous to the previously discovered α-glycerophosphate/dihydroxyacetone phosphate shuttle (e.g., Sacktor and Dick, 1962). In accord with the proposed function, it was shown that addition of catalytic quantities of Δ-pyrroline-5-carboxylate to a system reconstituted from soluble and mitochondrial components was capable of supporting sustained oxidation of NADH. The relative capacity for oxidation of proline and pyruvate by blowfly mitochondria (approximately 1 : 5, see Childress and Sacktor, 1966) would be in good accord with the postulated role, which would demand a ratio of not less than 1:6, and the existence of this mechanism would account for the otherwise puzzling observation that inhibition of the α-glycerophosphate shuttle system has no effect on the flight capacity of the blowfly (O'Brien *et al.*, 1965).

Further work is required to demonstrate the operation of this system *in vivo*, and until results become available it may be well to adhere to the view of Sacktor (1975) that "a complete explanation for the unique role of proline has yet to be formulated," noting in the meantime that the high energy yields which characterize its oxidation provides a supplementary source of energy during flight (Mayer and Candy, 1969), whatever subsidiary functions may be attributed to it.

Patterns of flight metabolism in different insects have been schematized in Fig. 3 to illustrate their differences. Figure 3a shows the situation in insects like the blowfly, where the metabolic pattern is dominated by the total oxidation of glucose, and proline metabolism is a minor element; in some beetles the "malic loop" has been more fully developed, to enable complete oxidation of proline as a supplementary source of energy for flight (Fig. 3b); whereas in the tsetse fly, and the Colorado beetle, where proline serves as the main source of energy for flight, the loop is used primarily as a source of amino acceptor for transamination, to produce a partially oxidized flight substrate (Fig. 3c). All these are variations on a basic theme, and the differences between them appear to result from quantitative modulations of a common enzyme pattern, which

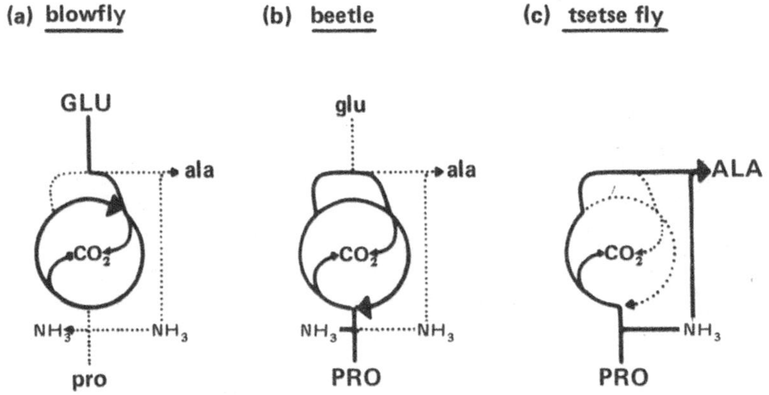

Fig. 3. Schematized comparison of flight metabolism in: the blowfly (*Phormia*), the Japanese beetle (*Popillia*), and the tsetse fly (*Glossina*). Main pathways are shown in heavy, and subsidiary pathways in dotted, lines. glu: glucose; ala:alanine; pro: proline.

affect enzymes of the glycolytic pathway as well as the key enzymes of proline metabolism, i.e., proline dehydrogenase, "malic enzyme," and alanine-oxoglutarate transaminase.

2.3. The Regulation of Proline Oxidation

During the transition from rest to flight the rate of proline oxidation increases enormously, and a number of regulatory effects have been thought to play a part. Regulation at the level of the respiratory chain is obviously of major importance in the tsetse fly, where respiratory control is tight, with high affinities for ADP and inorganic phosphate, and where ADP-stimulated respiration is completely sensitive to oligomycin (Bursell and Slack, 1976). In blowfly sarcosomes, a secondary requirement for much higher concentrations of ADP and inorganic phosphate is attributable to effects at the level of the dehydrogenases (Hansford and Sacktor, 1970). Additional control sites have been postulated on the basis of effects on glutamic dehydrogenase (Hansford and Johnson, 1975), which is potentiated by ADP and strongly inhibited by ammonia (Bond and Sang, 1968; Donnelan *et al.*, 1974; Bursell, 1975a; Khan and de Kort, 1978); on isocitrate dehydrogenase (Hansford and Johnson, 1975); and on proline dehydrogenase, which is inhibited by glutamate (Norden and Venturas, 1972) and which may be involved in substrate cycling (Crabtree and Newsholme, 1975). "Malic" enzyme from sarcosomes of the tsetse fly is strongly inhibited by ATP (Hansford and Johnson, 1975) and ADP (Norden and Matanganyidze, 1979), but the physiological significance of these effects has been questioned.

A general difficulty is inevitably involved in attempts to identify regulatory effects in a compartmentalized system that relies for its operation on the maintenance of steep concentration gradients between different parts—for ATP between mitochondrial surface and contractile proteins, for ADP and inorganic phosphate between contractile proteins and mitochondrial surface, for proline between hemolymph and mitochondrial surface, for ammonia between mitochondrion and hemolymph. Under these conditions, it is seldom possible to specify the concentration of supposed regulatory agents at the site of action. The best that can be done is to monitor shifts in average concentration during transitions from the resting to the active state (Sacktor and Hurlblut, 1966; Worm and Beenakkers, 1980), but these cannot be expected to provide a fully adequate basis for interpretation. While it may be accepted that control is exercised at the level of the respiratory chain, the possible importance of secondary effects must therefore remain an open question.

2.4. The Reconstitution of Proline

Insects generally appear to have a well developed capacity for proline synthesis, since proline always becomes strongly labeled following injection of a variety of radioactive substrates, including glucose (Kastings and McGinnis, 1958; Moloo *et al.*, 1974; Moloo 1976b), acetate (Winteringham and Harrison, 1956; Price, 1961), formate (McEnroe and Forgash, 1958), and a variety of amino acids (McCabe and Bursell, 1975a; Collett, 1976b; de Kort and Kramer, 1976; Hargrove, 1976; Moloo, 1976a and 1977a). Synthetic activity has been shown to be associated with the fat body (Clements, 1959), and metabolic pathways have been investigated with fat body isolated from the tsetse fly.

Alanine and triglyceride are the major substrates for proline synthesis in the tsetse fly (McCabe and Bursell, 1975b; Bursell, 1977), with other amino acids and glucose giving much lower rates of production. A transamination with oxoglutarate appears to be the first step, with pyruvate and glutamate as the products. The glutamate is reduced to proline, while the pyruvate is carboxylated to give oxaloacetate; this condenses with acetyl CoA from the β-oxidation of lipids, and the resulting citrate is metabolized through normal Krebs cycle intermediates to replace the oxoglutarate used in the initial transamination.

Since alanine is the main end product of proline oxidation, its use as a major substrate for proline synthesis establishes a close relationship between flight muscle and fat body, as illustrated in Fig. 4. Energy for flight is generated by the partial oxidation of proline, and the alanine so formed passes into the hemolymph for transport to the fat body. There it couples with two-carbon lipid fragments to reconstitute proline, which

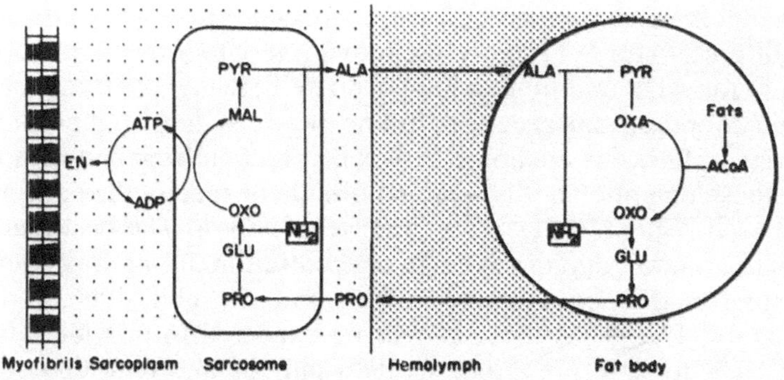

Fig. 4. Diagrammatic representation of the relationship between flight muscle and fat body metabolism in the tsetse fly (*Glossina*). Metabolites are denoted by the first three letters of their names. EN: energy; ACoA: acetyl CoA. From Bursell, 1977.

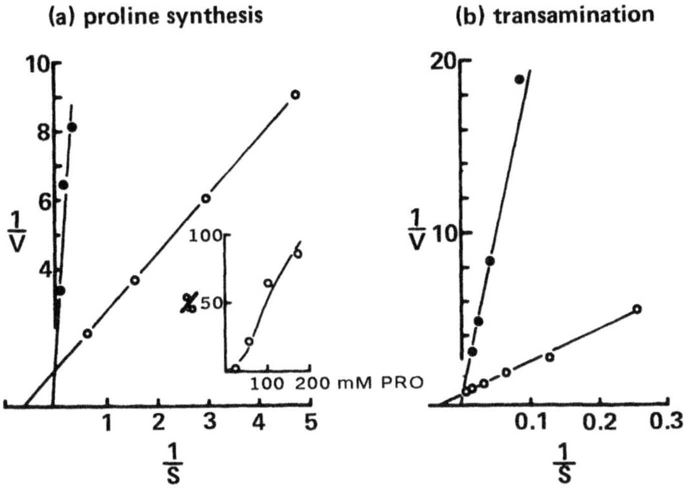

Fig. 5. The effect of proline on proline synthesis by fat body of the tsetse fly. a: Reciprocal plots of the relationship between L-alanine concentration (S) and the velocity of proline synthesis in the absence (○) and presence (●) of 100 mM L-proline. Inset shows the percentage inhibition (%) of proline synthesis as a function of proline concentration at a substrate concentration of 8.5 mM. All points represent the averages of 3–5 determinations made according to the methods described in Bursell (1977). b: Reciprocal plots of the relationship between L-alanine concentration (S) and the velocity of the reaction catalyzed by alanine-oxoglutarate aminotransferase in the absence (○) and presence (●) of 125 mM L-proline. The enzyme was extracted in 80 mM phosphate buffer (pH 7.4) from acetone powders of fat body from 2-week-old virgin females, and assays were made by the method described in Bergmeyer and Bernt, 1963.

passes into the hemolymph for transport back to the flight muscle. In its cyclic operation the sole input to the system would be lipid hydrocarbon, while the output would be energy and carbon dioxide, and the system is seen to provide a special mechanism for the oxidation of lipid geared to the special proline-centered metabolism of the tsetse fly. Alanine could, in a sense, be seen as a carrier of lipid between fat body and flight muscle.

In fact, the capacity of the fat body for proline synthesis is not commensurate with the capacity of the flight muscle for proline oxidation, so the system is not capable of steady state operation. The proline reserves would be supplemented to some extent by proline synthesis during flight (Bursell, 1978), but in the main, the oxidative part of the cycle predominates in the active phase, while the synthetic part predominates in the resting phase.

The operation of the system is subject to feedback control through an effect of proline on proline synthesis, as illustrated in Fig. 5. The affinity of the system for proline is extremely low, and it is not until proline concentrations reach about 150 mM (equal to the hemolymph concentration in resting flies) that the synthesis is fully inhibited, as shown by the inset in Fig. 5a. Hydroxyproline is also capable of inhibiting proline synthesis (Rapport *et al.*, 1979), but it is much less effective than proline itself. The low apparent K_m for the synthetic process (1.6 mM) is of interest, since the first step is thought to be catalyzed by alanine-oxoglutarate aminotransferase. This enzyme is also inhibited by proline (Fig. 5b), but in the tsetse fly, as in animals generally (e.g., Barman, 1969), it has a very much higher apparent K_m (25 mM). The possibility that the discrepancy could be accounted for on the basis of an active uptake of alanine by the fat body has been investigated, but no evidence of uptake was found.

3. Discussion

Accepting the importance of proline in the flight metabolism of many insects, and perhaps particularly of bloodsucking species, the question has been raised, why proline? (Kammer and Heinrich, 1978). It does not figure especially prominently in the composition of blood proteins, making up only 3.8% of the total, so why should it have been singled out from the rest of the amino acids as a flight substrate?

In considering this question, a number of factors would obviously need to be taken into account. To constitute a suitable flight substrate an amino acid would, in the first place, need to produce a high yield of metabolic energy. Secondly, it should be highly soluble, so that a large amount of energy could be held in solution in body fluids and, once tissue reserves have been depleted, so that steep concentration gradients could be established between hemolymph and flight muscle, thereby ensuring a rapid diffusion of material to the site of oxidation. Consideration would also need to be given to the amount of nitrogen that the molecule contains, since this would require detoxication by conversion to uric acid at a cost of 1.5 mol of ATP per atom of nitrogen. On this basis, a number of amino acids could immediately be ruled out as potential flight substrates. The proportionate nitrogen content of arginine and histidine, for example, rivals that of uric acid itself, and it would be metabolically uneconomical to deaminate these amino acids, since the slight metabolic gain which could be made from deaminated products would be offset by metabolic losses associated with the detoxication of their nitrogen. It is not surprising, therefore, that in the tsetse fly these two amino acids are quantitatively excreted (Bursell, 1965b; Moloo, 1977a). Other amino acids, such as tyrosine and cystine, are too sparingly soluble to warrant consideration as potential flight substrates, and this would leave the limited list shown in Table 2 for consideration. Their net energy yields, calculated

Table 2. The Energy Yield and Solubility of Potential Amino Acid Flight Substrates[a]

Amino acid	Net ATP Yield (mol/g)	Solubility mM (50% saturated)
Leucine	33.2	150
Isoleucine	32.4	150
Proline	27.6	>1000
Valine	24.2	385
Threonine	22.3	—
Alanine	18.5	930
Glutamate	17.3	30
Lysine	17.1	>1000
Glutamine	16.4	145
Serine	15.7	>1000
Glycine	12.7	>1000
Aspartate	12.4	31
Asparagine	11.4	93

[a] Solubility data from Hodgman *et al.*, 1957.

as the amount of ATP which can be generated by total oxidation less the amount required to detoxicate their nitrogen, are listed in column 2, and it can be seen that using this criterion the only serious rivals to proline would be leucine and isoleucine. However, as shown in column 3, their solubility is relatively low, and the concentration at 50% saturation would be well below recorded levels of proline concentration in the general tissue fluids of the tsetse fly, which exceed 210 mM (from data in Bursell, 1973, and Hargrove, 1975).

In view of these considerations, it could be argued that, of all the amino acids that make up the blood proteins, proline would be the most suitable as a flight substrate, but it could still be asked why an amino acid should be used as a flight substrate instead of converting it to carbohydrate, which combines high solubility with reasonable energy yield and in addition has the potential for convenient storage, which is so notably lacking in amino acids. In this context it should be noted, however, that the requirement for disposal of amino nitrogen in the form of uric acid is not easily reconciled with the conversion of protein nutriment to carbohydrate. This is because the main uric acid precursor, glycine, plus its metabolically near relatives alanine and serine, are the very amino acids which could most easily be converted to 3-carbon intermediates of the gluconeogenic pathway, but nearly 60% of the total intake of these three amino acids is required to dispose, as uric acid, of the protein nitrogen with which they are associated (from McCabe and Bursell, 1975a). Other amino acids, which degrade predominantly to 4-carbon or 2-carbon fragments could not economically be converted to 3-carbon fragments in view of the irreversibility of the reaction which converts pyruvate to acetyl CoA.

These considerations go some way toward establishing the suitability of proline as a substrate of flight metabolism in insects generally, and in bloodsucking species in particular, among which it may not be too far-fetched to see the adoption of a proline-based metabolism as a specific biochemical adaptation to the bloodsucking mode of life.

NOTE ADDED IN PROOF. Since the preparation of this contribution, reports have appeared on the investigation of proline synthesis by fat body of *Leptinotarsa decemlineata* (Weeda *et al.*, 1980b). The system appears to be very similar to the one described for the tsetse fly.

ACKNOWLEDGMENTS. My thanks are due to Dr. E.A. Newsholme for helpful comments on the manuscript.

References

Bailey, E., 1975, Biochemistry of insect flight: Part 2. Fuel supply, in: *Insect Biochemistry and Function* (D.J. Candy and B.A. Kilby, eds.), pp. 95–176, Chapman and Hall, London.

Balboni, E., 1978, A proline shuttle in insect flight muscle, *Biochem. Biophys. Res. Commun.* **85:**1090.

Barker, R.J., and Lehner, Y., 1972, Free amino acids in thoraces of flown honey bees, *Apis mellifera* L. (Hymenoptera: Apidae), *Comp. Biochem. Physiol.* **43B:**163.

Barman, T.E., 1969, *Enzyme Handbook*, Volume 1, p. 357, Springer Verlag, Berlin, Heidelberg, New York.

Barrett, F.M., 1974, Changes in the concentration of free amino acids in the haemolymph of *Rhodnius prolixus* during the fifth instar, *Comp. Biochem. Physiol.* **48B:**241.

Barrett, R.M., and Friend, W.G., 1975, Differences in the concentration of free amino acids in the haemolymph of adult male and female *Rhodnius prolixus, Comp. Biochem. Physiol.* **52B:**427.

Bergmeyer, H.V., and Bernt, E., 1963, Glutamate-Pyruvate Transaminase: UV-assay, manual method, in: *Methods in Enzymatic Analysis* (H.V. Bergmeyer, ed.), pp. 752–758, Academic Press, New York, London.

Birtwisle, D., 1974, Sugar phosphates and amino-acid concentrations in the thorax of the tsetse fly *Glossina morsitans, Insect Biochem.* **4:**63.

Bond, P.A., and Sang, J.H., 1968, Glutamate dehydrogenase in *Drosophila* larvae, *J. Insect Physiol.* **14:**341.

Brady, J., 1970, Characteristics of spontaneous activity in tsetse flies, *Nature (London)* **228:**286.

Brosemer, R.W., and Veerabhadrappa, P.S., 1965, Pathways of proline oxidation in insect flight muscle, *Biochim. Biophys. Acta* **110:**102.

Brouwers, E. V.M., and de Kort, C.A.D., 1979, Amino acid metabolism during flight in the Colorado beetle, *Leptinotarsa decemlineata* Say., *J. Insect Physiol.* **25:**411.

Bursell, E., 1963, Aspects of the metabolism of amino acids in the tsetse fly *Glossina* (Diptera), *J. Insect Physiol.* **9:**439.

Bursell, E., 1965a, Oxaloacetic decarboxylase in flight musculature of the tsetse fly, *Comp. Biochem. Physiol.* **16**:259.

Bursell, E., 1965b, Nitrogenous waste products of the tsetse fly *Glossina morsitans*, *J. Insect Physiol.* **11**:993.

Bursell. E., 1966, Aspects of the flight metabolism of tsetse flies (*Glossina*), *Comp. Biochem. Physiol.* **19**:809.

Bursell, E., 1967, The conversion of glutamate to alanine in the tsetse fly (*Glossina morsitans*), *Comp. Biochem. Physiol.* **23**:825.

Bursell, E., 1973, Development of mitochondrial and contractile components of the flight muscle in adult tsetse flies, *Glossina morsitans*, *J. Insect Physiol.* **19**:1079.

Bursell, E., 1975a, Glutamic dehydrogenase from sarcosomes of the tsetse fly (*Glossina morsitans*) and the blowfly (*Sarcophaga nodosa*), *Insect Biochem.* **5**:289.

Bursell, E., 1975b, Substrates of oxidative metabolism in Dipteran flight muscle, *Comp. Biochem. Physiol.* **52B**:235.

Bursell, E., 1977, Synthesis of proline by fat body of the tsetse fly (*Glossina morsitans*): Metabolic pathways, *Insect Biochem.* **7**:427.

Bursell, E., 1978, Quantitative aspects of proline utilisation during flight in tsetse flies, *Physiol. Entomol.* **3**:265.

Bursell, E., and Slack, E., 1976, Oxidation of proline by sarcosomes of the tsetse fly, *Glossina morsitans*, *Insect Biochem.* **6**:159.

Bursell, E., Billing, K.C., Hargrove, J.W., McCabe, C.T., and Slack, E., 1974, Metabolism of the bloodmeal in tsetse flies: A review, *Acta Trop.* **31**:297.

Chen, P.S., 1966, Amino acid and protein metabolism in insect development, *Adv. Insect Physiol.* **3**:53.

Childress, C.C., and Sacktor, B., 1966, Pyruvate oxidation and the permeability of mitochondria from blowfly flight muscle, *Science* **154**:268.

Clements, A.N., 1959, Studies on the metabolism of locust fat body, *J. Exp. Biol.* **36**:665.

Collett, J.I., 1976a, The constancy and similarity of the amounts of free amino acids in inbred strains of *Drosophila* and outbred *Calliphora*, *J. Insect Physiol.* **22**:1251.

Collett, J.I., 1976b, Some features of the regulation of the free amino acids in adult *Calliphora erythrocephala*, *J. Insect Physiol.* **22**:1395.

Corrigan, J.J., and Kearns, C.W., 1963, Amino acid metabolism in DDT-poisoned American cockroaches, *J. Insect Physiol.* **9**:1.

Crabtree, B., and Newsholme, E.A., 1970, The activities of proline dehydrogenase, glutamate dehydrogenase, aspartate-oxoglutarate aminotransferase and alanine oxoglutarate aminotransferase in some insect flight muscles, *Biochem. J.* **117**:1019.

Crabtree, B., and Newsholme, E.A., 1975, Comparative aspects of fuel utilization and metabolism by muscles, in: *Insect Muscle* (P.N.R. Underwood, ed.), pp. 405–501, Academic Press, London.

D'Costa, M.A., Rice, M.J., and Latif, A., 1973, Glycogen in the proventriculus of the tsetse fly, *J. Insect Physiol.* **19**:427.

de Kort, C.A.D., and Kramer, S.J., 1976, Age-related changes in amino acid composition and turnover in the Colorado potato beetle, *Leptinotarsa decemlineata*, *Insect Biochem.* **6**:339.

de Kort, C.A.D., Bartelink, A.K.M., and Schuurmans, R.R., 1973, The significance of L-proline for oxidative metabolism in the flight muscles of Colorado beetle, *Leptinotarsa decemlineata*, *Insect Biochem.* **3**:11.

Donnelan, J.F., Jenner, D.W., and Ramsey, A., 1974, Subcellular fractionation of fleshfly flight muscle in an attempt to isolate synaptosomes and to establish the location of glutamate enzymes, *Insect Biochem.* **4**:243.

Florkin, M., 1958, The free amino acids of insect haemolymph, *Proc. 4th Int. Congr. Biochem., Vienna* **12:**63.

Geigy, R., Huber, M., Weinman, D., and Wyatt, G.R., 1959, Demonstration of trehalose in the vector of trypanosomiasis: The tsetse fly, *Acta Trop.* **16:**255.

Hansford, R.G., and Johnson, R.N., 1975, The nature and control of the tricarboxylate cycle in beetle flight muscle, *Biochem. J.* **148:**389.

Hansford, R.G., and Sacktor, B., 1970, The control of oxidation of proline by isolated flight muscle mitochondria. *J. Biol. Chem.* **245:**991.

Hargrove, J.W. 1975, Some changes in the flight apparatus of tsetse flies, *Glossina morsitans* and *G. pallidipes* during maturation, *J. Insect Physiol.* **21:**1485.

Hargrove, J.W., 1976, Amino acid metabolism during flight in tsetse flies, *J. Insect Physiol.* **22:**309.

Hochachka, P.W., and Somero, A.N., 1973, *Strategies of Biochemical Adaptation* W.B. Saunders, Philadelphia, London, Toronto.

Hodgman, C.D., Weast, R.C., and Selby, S.M., 1957, *Handbook of Chemistry and Physics*, 39th edition, Chemical Rubber Publishing Company, Cleveland, Ohio.

Hoek, J.B., Pearson, D.J., and Olembo, N.K., 1976, Nicotinamide dinucleotide-linked 'malic' enzyme in flight muscle of the tsetse fly (*Glossina*) and other insects, *Biochem. J.* **160:**253.

Jeuniaux, C., 1971, Haemolymph-Arthropoda, in: *Chemical Zoology*, Volume VI (M. Florkin and B.T. Scheer, eds.,), pp. 64–118 Academic Press, New York and London.

Kammer, A.E., and Heinrich, B., 1978, Insect flight metabolism, *Adv. Insect Physiol.* **13:**133.

Kastings, R., and McGinnis, A.J., 1958, Use of glucose labelled with carbon-14 to determine the amino-acids essential for an insect, *Nature (London)* **182:**1380.

Khan, M.A., and de Kort, C.A.D., 1978, Further evidence for the significance of proline as a substrate for flight in the Colorado potato beetle, *Leptinotarsa decemlineata, Insect Biochem.* **3:**403.

Kirsten, E., Kirsten, R., and Arese, P., 1963, The content of free amino acids, energy-rich phosphoric acid and glycolytic and three-carbon acid substrates in the muscle of *Locusta migratoria* during work, *Biochem. Z.* **337:**167.

Lewis, D.L., 1976, Aspects of amino acid metabolism in *Stomoxys calcitrans*, M.Sc. thesis, University of Rhodesia.

Lewis, S.E., and Price, G.M., 1956, "Malic enzyme" activity in blowfly muscle, *Nature (London)* **177:**842.

McCabe, C.T., and Bursell, E., 1975a, The metabolism of digestive products in the tsetse fly, *Glossina morsitans, Insect Biochem.* **5:**769.

McCabe, C.T., and Bursell, E., 1975b, Interrelationships between amino acid and lipid metabolism in the tsetse fly, *Glossina morsitans, Insect Biochem.* **5:**781.

McEnroe, W.O. and Forgash, A.J., 1958, Formate metabolism in the American cockroach, *Periplaneta americana, Ann. Entomol. Soc. Am.* **51:**126.

Mayer, R.J., and Candy, D.J., 1969, Changes in energy reserves during flight in the desert locust, *Schistocerca gregaria, Comp. Biochem. Physiol.* **31:**409.

Moloo, S.K., 1976a, Nutrition of *Glossina morsitans*: Metabolism of U-^{14}C threonine during pregnancy, *Acta Trop.* **33:**133.

Moloo, S.K., 1976b, Nutrition of *Glossina morsitans*: Metabolism of U-^{14}C glucose during pregnancy, *J. Insect Physiol.* **22:**195.

Moloo, S.K., 1977a, Aspects of the metabolism of U-^{14}C arginine, U-^{14}C histidine and U-^{14}C lysine by adult female *Glossina morsitans* during pregnancy, *Comp. Physiol. Biochem.* **57B:**23.

Moloo, S.K., 1977b, Metabolism of U-^{14}C phenylalanine and U-^{14}C tyrosine by females of

Glossina morsitans Westw. (Diptera: Glossinidae) during pregnancy, *Bull. Entomol. Res.* **67**:651.

Moloo, S.K., Langley, P.A., and Balogun, R.A., 1974, Amino acid synthesis from glucose-U-^{14}C in *Glossina morsitans, J. Insect Physiol.* **20**:1807.

Mordue, W., and de Kort, C.A.D., 1978, Energy substrates for flight in the Colorado beetle, *Leptinotarsa decemlineata*, Say., *J. Insect Physiol.* **24**:221.

Muus, J., 1968, Gel electrophoresis of some dehydrogenases from flight muscles of tsetse flies (*Glossina morsitans*) and houseflies (*Musca domestica*), *Comp. Biochem. Physiol.* **24**:527.

Nayar, J.K., and van Handel, E., 1972, Utilisation of injected glucose by the tsetse fly (*Glossina*) and the stable fly (*Stomoxys*), *J. Insect Physiol.* **18**:105.

Norden, D.A., and Matanganyidze, C., 1977, Some properties of a mitochondrial malate enzyme from the flight muscle of the tsetse fly (*Glossina*), *Insect Biochem.* **7**:215.

Norden, D.A., and Matanganyidze, C., 1979, Activities of Krebs' cycle enzymes in the flight muscles of the tsetse fly (*Glossina*) and the blowfly (*Sarcophaga*), *Insect Biochem.* **9**:85.

Norden, D.A., and Paterson, D.J., 1969, Carbohydrate metabolism in flight muscles of the tsetse fly (*Glossina*) and the blowfly (*Sarcophaga*), *Comp. Biochem. Physiol.* **31**:819.

Norden, D.A., and Paterson, D.J., 1970, Carbohydrate metabolism in flight muscles of the tsetse fly (*Glossina*) and the blowfly (Sarcophaga): II. *Int. J. Biochem.* **1**:81.

Norden, D.A., and Venturas, D., 1972, Substances affecting the activity of proline dehydrogenase in the sarcosomes of the tsetse fly (*Glossina*) and a comparison with some other insects, *Insect Biochem.* **2**:226.

Nurmi, S., and Birt, L.M., 1974, Amino-acid pools in developing adults of the blowfly *Lucilia cuprina* after the pupal-adult ecdysis, *Insect Biochem.* **4**:287.

O'Brien, R.D., Cheung, L., and Kimmel, E., 1965, Inhibition of the α-glycerophosphate shuttle in housefly flight muscle, *J. Insect Physiol.* **11**:1241.

Price, G.M., 1961, Some aspects of amino acid metabolism in the adult housefly, *Musca domestica, Biochem. J.* **80**:420.

Rapport, E.W., Yang, M.K., and Tobe, S.S., 1979, Hydroxyproline as an inhibitor of proline synthesis in *Drosophila melanogaster, Insect Biochem.* **9**:19.

Ray, J.W., 1964, The free amino acid pool of the cockroach (*Periplaneta americana*) central nervous system, and the effect of insecticides, *J. Insect Physiol.* **10**:587.

Reddy, S.R.R., and Campbell, J.W., 1969, Arginine metabolism in insects: Role of arginase in proline formation during pupal development, *Biochem J.* **115**:495.

Sacktor, B., 1961, The role of mitochondria in respiratory metabolism in flight muscle, *Annu. Rev. Entomol.* **6**:103.

Sacktor, B., 1970, Regulation of intermediary metabolism with special reference to the control mechanisms in insect flight muscle, *Adv. Insect Physiol.* **7**:286.

Sacktor, B., 1974, Biological oxidations and energetics in insect mitochondria, in: *The Physiology of Insecta*, Volume IV, 2nd edition (M. Rockstein, ed.), pp. 271–354, Academic Press, New York and London.

Sacktor, B., 1975, Biochemistry of insect flight. Part I. Utilization of fuels by muscle, in: *Insect Biochemistry and Function* (D.J. Candy and B.A. Kilby, eds.), pp. 1–88, Wiley, New York.

Sacktor, B., and Childress, C.C., 1967, Metabolism of proline in insect flight muscle and its significance in stimulating the oxidation of pyruvate, *Arch. Biochem. Biophys.* **120**:583.

Sacktor, B., and Dick, A.R., 1962, Pathways of hydrogen transport in the oxidation of extramitochondrial reduced diphosphopyridine nucleotide in flight muscle, *J. Biol. Chem.* **237**:3259.

Sacktor, B., and Hurlblut, E.C., 1966, Regulation of metabolism in working muscle *in vivo*. II. Concentration of adenine nucleotides, arginine phosphate and inorganic phosphate in insect muscle during flight, *J. Biol. Chem.* **241**:632.

Sacktor, B., and Wormser-Shavit, E., 1966, Regulation of metabolism in working muscle *in vivo*. I. Concentration of some glycolytic, tricarboxylic acid and amino acid intermediates in insect flight muscle during flight, *J. Biol. Chem.* **241**:629.

Sacktor, B., Childress, C.C., and Wormser-Shavit, E., 1965, A special role for proline in flight muscle, *Fed. Proc. Fed. Am. Soc. Exp. Biol.* **24**:271.

Schoffeniels, E., and Gilles, R., 1970, Nitrogenous constituents and nitrogen metabolism in arthropods, in: *Chemical Zoology*, Volume VA (M. Florkin and B.T. Scheer, eds.), pp. 199–227, Academic Press, New York and London.

Slack, E., and Bursell, E., 1976, Oxidation of pyruvate by mitochondria isolated from flight muscles of blowflies, *Insect Biochem.* **6**:637.

Stevenson, E., 1968, Carbohydrate metabolism in the flight muscles of the Southern Army Moth, *Prodenia eridania*, *J. Insect Physiol.* **14**:179.

Sutcliffe, D.W., 1963, The chemical composition of haemolymph in insects and some other arthropods in relation to their phylogeny, *Comp. Biochem. Physiol.* **9**:121.

van den Bergh, S.G., 1964, Pyruvate oxidation and the permeability of housefly sarcosomes, *Biochem. J.* **93**:128.

Wadano, A., Yamamoto, T., Yoshida, K., and Murra, K., 1976, Pyrroline-carboxylate reductase of a blowfly, *Aldrichina grahami*, *Insect Biochem.* **6**:657.

Weeda, E., de Kort, C.A.D., and Beenakkers, A.M.T., 1979, Fuels for energy metabolism in the Colorado beetle, *Leptinotarsa decemlineata* Say., *J. Insect Physiol.* **25**:951.

Weeda, E., de Kort, C.A.D., and Beenakkers, A.M.T., 1980a, Oxidation of proline and pyruvate in flight muscle mitochondria of the Colorado beetle *Leptinotarsa decemlineata* Say., *Insect Biochem.* **10**:305.

Weeda, E., Koopmanschap, A.B., de Kort, C.A.D., and Beenakkers, A.M.T., 1980b, Proline synthesis in fat body of *Leptinotarsa decemlineata*, *Insect Biochem.* **10**:631.

Winteringham, F.P.W., 1958, Comparative aspects of insect biochemistry with particular reference to insecticidal action, *Proc. 4th Int. Congr. Biochem., Vienna* **12**:201.

Winteringham, F.P.W., and Harrison, A.H., 1956, Study of anticholinesterase action in insects by a labelled pool technique, *Nature (London)* **178**:81.

Worm, R.A.A., and Beenakkers, A.M.T., 1980, Regulation of substrate utilisation in the flight muscle of the locust *Locusta migratoria* during flight, *Insect Biochem.* **10**:53.

Wyatt, G.R., 1961, The biochemistry of insect haemolymph, *Annu. Rev. Entomol.* **6**:75.

Yoshida, K., Wadano, A., and Murra, K., 1977, Metabolism of proline and its physiological significance in the larval blowfly *Aldrichina grahami*, *Insect Biochem.* **7**:51.

6

Lipid Transport by Hemolymph Lipoprotein

A Possible Multiple Role of Diacylglycerol-Carrying Lipoprotein

Haruo Chino

1. Introduction

The predominant form in which lipid is transported from fat body is as diacylglycerol associated with a specific hemolymph lipoprotein. The initial reports for the silkworm, *Hyalophora cecropia*, the grasshopper, *Melanoplus differentialis*, and the cockroach, *Periplaneta americana*, (Chino and Gilbert, 1964, 1965) have been confirmed by many investigators and extended to other species, including the locust, *Locusta migratoria* (Tietz, 1967).

The specific hemolymph lipoprotein was first isolated and purified from the silkworms *Philosamia cynthia* (Chino *et al.*, 1969) and *Hyalophora cecropia* (Thomas and Gilbert, 1968) and named "diacylglycerol-carrying lipoprotein" (DGLP) (equivalent to high density lipoprotein). Chino *et al.* (1969) have demonstrated *in vitro* the capacity of DGLP purified from *P. cynthia* to specifically take up diacylglycerol from fat body and have extensively characterized the molecular nature of the DGLP. Isolation and purification of DGLP from locust hemolymph was first reported by Peled and Tietz (1973, 1975) and more recently by

Haruo Chino • Biochemical Laboratory, Institute of Low Temperature Science, Hokkaido University, Sapporo, Japan.

Mwangi and Goldsworthy (1977) and Gellissen and Emmerich (1980), with the latter workers also characterizing the molecule.

Previous studies on lipid transport in insects have concentrated on the transport of lipid from the storage site in fat body, but few data are available to describe lipid transport from the site of lipid absorption in the midgut. Recently, we have demonstrated that in the American cockroach, diacylglycerol is the major form in which lipid is transported from the fat body and from the midgut (Chino and Downer, 1979).

The present chapter describes the purification of DGLP from hemolymph of the American cockroach and demonstrates the capacity of DGLP to accept diacylglycerol from fat body and midgut. The report also indicates the presence of a considerable amount of hydrocarbon in DGLP in addition to diacylglycerol and cholesterol. The possibility that, in the cockroach, DGLP serves to transport several different lipid classes will be discussed, and this possibility will also be considered for other insect species.

2. Purification and Function of DGLP

A simple and efficient method for purifying DGLP from pupal hemolymph of *P. cynthia* was reported by Chino *et al.* (1969). The method includes a specific precipitation at low ionic concentration at about pH 6.5, followed by column chromatography on DEAE-cellulose. The highly efficient precipitation step involves coprecipitation of two DGLPs (DGLP-I and DGLP-II). DGLP-I accepts diacylglycerol from incubating fat body, whereas DGLP-II does not have this capacity and, indeed, has been identified as the female-specific protein, vitellogenin (Chino *et al.*, 1976). Identification of the coprecipitant DGLP-II as vitellogenin indicates that the procedure for purifying DGLP-I is applicable only to hemolymph from female silkworms. In terms of function, DGLP-I is the only true diacylglycerol-carrier that has been identified; therefore, in the present account, the term DGLP will be restricted to DGLP-I.

The purification procedure developed for *C. cynthia* has been successfully adapted for the purification of DGLP and/or vitellogenin of other insects. Chinzei *et al.* (1981) used a slightly modified version of the procedure to obtain highly purified vitellogenin from hemolymph of adult locusts, and recently the method has been used to purify DGLP from hemolymph of the cockroach, *P. americana* (Chino *et al.*, 1981). The DGLP of cockroach resembles that of the silkworm in that it precipitates with a coprecipitant(s) under low ionic concentration at pH 6.0. However, in the cockroach, the coprecipitant may not be vitellogenin because coprecipitation occurs in the purification of DGLP from male cockroaches; thus, the purification procedure is applicable for both sexes of *P. amer-*

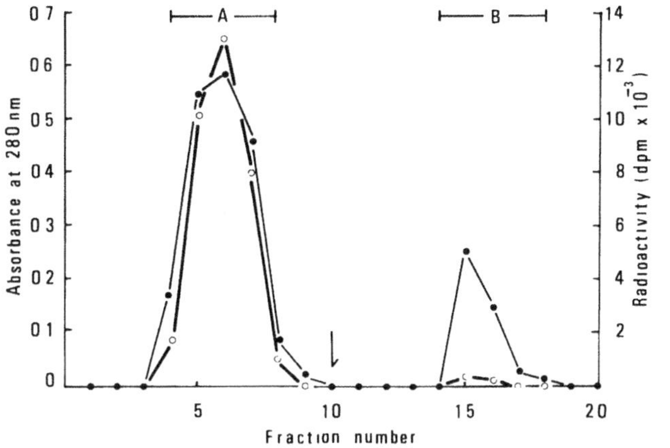

Fig. 1. DEAE-cellulose column chromatography of the fraction precipitated under low ionic concentration at pH 6.0. Original labeled hemolymph was collected and pooled from 10 female cockroaches 4 hr after injection of [1-^{14}C]palmitic acid (approximately 1 × 10^6 cpm/animal). One milliter fractions were collected in each tube. Arrow: beginning of elution with 0.25 M KCl in 0.05 M phosphate buffer, pH 6.0. Open circles: radioactivity, solid circles: absorbance at 280 nm. From Chino *et al.*, 1981.

icana. The mechanism of coprecipitation has not yet been elucidated.

The precipitate of cockroach DGLP and coprecipitant(s) is collected by centrifugation, redissolved in 0.05 M phosphate buffer (pH 6.0) and applied to a DEAE-cellulose column. A typical elution profile obtained from the hemolymph of female cockroaches that had been previously injected with [1-^{14}C]palmitic acid is illustrated in Fig. 1, and demonstrates that DGLP (fraction A) passes directly through the column, whereas a second protein fraction (fraction B) is eluted subsequently with 0.25 M KCl. A similar elution profile is obtained from hemolymph of male cockroaches. The homogeneity of the DGLP preparation is confirmed by polyacrylamide gel electrophoresis (Figs. 2a,b).

The method employed to purify DGLP from cockroach hemolymph may also be applied to the purification of DGLP from male and female adult locusts. Figures 2c and 2d demonstrate the homogeneity of the DGLP preparation from locusts.

When cockroaches are injected with or fed [1-^{14}C]palmitic acid prior to collection of hemolymph, most of the radioactivity associated with purified DGLP is found in the diacylglycerol fraction (Table 1). In addition, the data presented in Table 2 and 3 demonstrate the capacity of purified DGLP to take up diacylglycerol from the fat body and midgut *in vitro*. These observations suggest that the same molecule, DGLP, serves to transport diacylglycerol from two different sites, the storage site in the fat body and the absorption site in midgut.

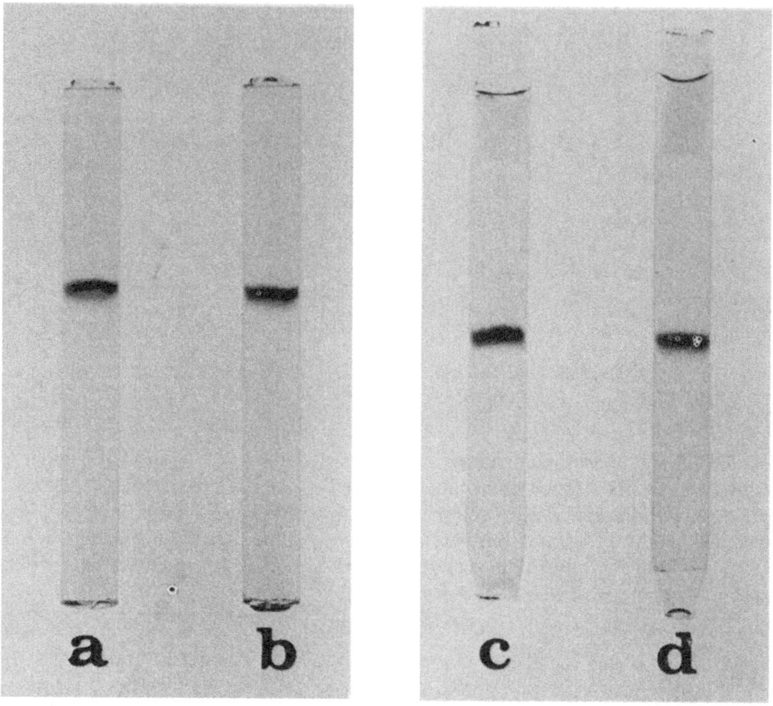

Fig. 2. Polyacrylamide gel electrophoresis of DGLP purified from hemolymph of adult American cockroach and locust. a: Preparation from female cockroach. b: Preparation from male cockroach. c: Preparation from female locust. d: Preparation from male locust. (a and b from Chino *et al.*, 1981.)

Table 1. Distribution of Radioactivity in Lipids Associated with Cockroach DGLP[a,b]

	Prelabeled by injection		Prelabeled by feeding	
Lipid class	Radioactivity (dpm)	Percentage	Radioactivity (dpm)	Percentage
Total lipids	90,440	100	577,910	100
1. Hydrocarbon	16,520	18.3	1,790	0.3
2. Cholesterol ester	357	0.4	Insignificant	—
3. Triacylglycerol	1,120	1.2	468	0.1
4. Diacylglycerol	67,390	74.5	543,090	94.0
5. Monoacylglycerol	Insignificant	—	2,990	0.5
6. Free fatty acid	1,030	1.1	1,300	0.2
Sum of 1–6	86,417	95.5	549,638	95.1

[a] DGLP was purified from the pooled hemolymph of 10 female cockroaches that received approximately 1×10^6 cpm [1-^{14}C]palmitic acid/animal by injection or feeding. When the insects feed on [1-^{14}C]palmitic acid, the labeled fatty acids are first taken up by intestinal cells and undergo the esterification via the monoacylglycerol acyl transferase system to diacylglycerol (Hoffman and Downer, 1979), and the labeled diacylglycerol is subsequently released into hemolymph.
[b] From Chino *et al.*, 1981.

Table 2. Assay of Uptake of Diacylglycerol by Cockroach DGLP from Prelabeled Fat Body[a]

Exp. No.[b]	Incubation medium	Protein amount used for assay (mg)	[^{14}C]Diacylglycerol released into incubation medium (dpm)
1	Phosphate saline	0	4,890[c]
	Purified DGLP	0.60	27,110
	Fraction B	0.60	4,630
2	Phosphate saline	0	1,080
	Fresh hemolymph	0.60	4,600
	Purified DGLP	0.43	12,780
	Fraction B	0.43	1,440

[a] From Chino *et al.*, 1981.
[b] Experiment 1: 150 mg prelabeled fat body containing 150,200 dpm [^{14}C]diacylglycerol in each incubation; Experiment 2: 150 mg prelabeled fat body containing 53,900 dpm [^{14}C]diacylglycerol. Incubation time: 90 min at 25°C. For definition of fraction B, see Fig. 1.
[c] Release into phosphate saline represents only "artificial leaking" due to disintegration of prelabeled fat body during incubation.

Table 3. Assay of Uptake of Diacylglycerol by Cockroach DGLP from Isolated Prelabeled Midgut[a,b]

Incubation medium	Protein amount used for assay (mg)	[^{14}C]Diacylglycerol released into incubation medium (dpm)
Phosphate saline	0	1,810
Phosphate saline	0	1,730
Phosphate saline	0	1,070
Fresh hemolymph	1.8	11,210
Fresh hemolymph	1.0	4,240
Purified DGLP	1.0	9,350
Purified DGLP	0.46	6,060
Fraction B	0.34	1,490
Fraction B	0.40	864

[a] Incubation time: 90 min at 25°C. Amounts of [^{14}C]triacylglycerol and [^{14}C]palmitic acid released into incubation media were negligible.
[b] From Chino *et al.*, 1981.

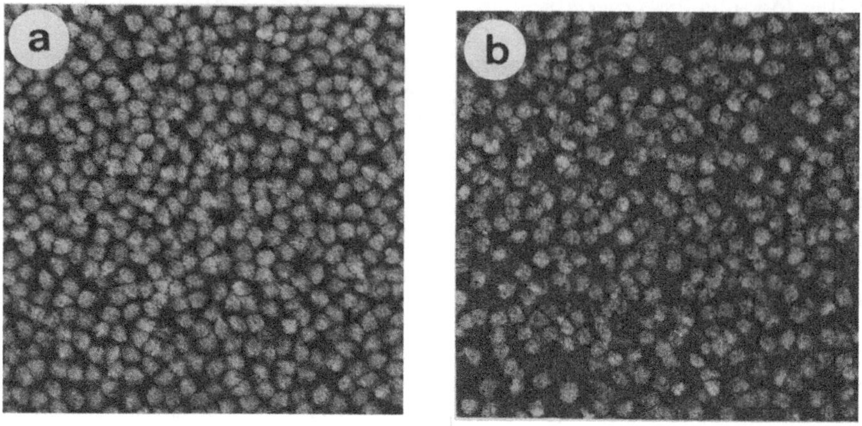

Fig. 3. Electron micrographs of DGLPs negatively stained with uranium acetate (×150,-000). (a): DGLP purified from cockroach hemolymph. (b): DGLP purified from hemolymph of *Philosamia* silkworm.

3. Physicochemical Nature of DGLP

Negatively stained electron micrographs of cockroach and silkworm DGLPs are illustrated in Fig. 3 and further demonstrate the homogeneous nature of the DGLP preparation. The cockroach DGLP is similar in shape and size to the silkworm molecule.

Some physical and chemical properties of cockroach DGLP are presented in Table 4 and compared with those reported previously for DGLPs of *P. cynthia* (Chino *et al.*, 1969) and *L. migratoria* (Gellissen and Em-

Table 4. Some Physicochemical Characteristics of DGLPs from Different Insects

	Philosamia silkworm[a]	American cockroach[b]	Locust[c]
Molecular weight (by sedimentation-equilibrium)	700,000	600,000	850,000
Molecular shape	Almost globular	Almost globular	Unknown
Molecular size	130 ± 5 Å	160 ± 11 Å	Unknown
Color	Deep yellow	Pale yellow	Yellow
Total lipid content	45%	50%	36%
Mannose content	1.2%	0.9%	1%
Glucosamine content	0.5%	0.3%	—
Subunits (molecular weight)	Heavy chain (270,000) Light chain (88,000)	Heavy chain (250,000) Light chain (85,000)	Single (85,000)

[a] Mostly from Chino *et al.*, 1969.
[b] From Chino *et al.*, 1981.
[c] From Gellissen and Emmerich, 1980.

Table 5. Amino Acid Composition of DGLP from Different Insects

	Recovered amino acids (mol/1000 mol)		
Amino acid	Cockroach[a]	Silkworm[b]	Locust[c]
Asp	110	126	119
Thr	66	49	55
Ser	69	69	76
Glu	108	104	120
Pro	38	47	51
Gly	64	67	62
Ala	68	63	70
Val	84	74	70
Met	3	5	—
Ile	41	58	51
Leu	107	90	96
Tyr	30	28	39
Phe	47	48	47
His	39	28	19
Lys	93	107	76
Arg	27	37	39
Cys	6	—	—

[a] From Chino *et al.*, 1981.
[b] From Chino *et al.*, 1969.
[c] From Gellissen and Emmerich, 1980.

Table 6. Lipid Composition of DGLPs from American Cockroach and *Philosamia* Silkworm

	Content (%)	
Component	Cockroach[a]	Silkworm[b]
Protein	50.4	56.0
Total lipids	49.6	44.0
1. Hydrocarbon[c]	28.3	Insignificant
2. Triacylglycerol	2.0	1.2
3. Diacylglycerol	15.2	56.3
4. Monoacylglycerol	0.0	0.0
5. Cholesterol	5.0	13.2
6. Cholesterol ester	0.0	0.0
7. Total phospholipids	42.8	25.8
Phosphatidylcholine	(68)[d]	(48)
Phosphatidylethanolamine	(32)	(32)
Sphingomyelin	(0)	(20)
8. Sum of 1–7	93.4	95.5

[a] From Chino *et al.*, 1981.
[b] From Chino *et al.*, 1969.
[c] Lipid fractions (1–8) are expressed as percentages of the total lipids.
[d] Figures in parentheses indicate percentage of total phospholipids.

Table 7. Hydrocarbon Composition of Cockroach DGLP[a]

Hydrocarbons	Content (%)
1. *n*-Pentacosane	7.2
2. 3-Methylpentacosane	18.1
3. 6,9-Heptacosadiene	72.3
4. Unidentified hydrocarbon	2.4

[a] From Chino *et al.*, 1981.

merich, 1980). The three DGLPs appear to differ only in the subunit structures of the apoproteins. Further similarity between the three DGLPs is indicated by comparison of the amino acid compositions (Table 5); all three DGLPs are rich in glutamic acid and aspartic acid, but extremely low in methionine. The lipid compositions of cockroach and silkworm DGLP are indicated in Table 6 (no reliable data are available for locust DGLP). Of particular interest is the presence, in cockroach DGLP, of a considerable amount of hydrocarbon in addition to diacylglycerol. This finding supports the earlier observation that significant radioactivity is recovered from the hydrocarbon fraction of DGLP, when [1-^{14}C]palmitic acid is injected into the cockroach hemocoel (Table 1). Previous studies have reported that the carboxyl carbon of long chain fatty acid is incorporated into hydrocarbon by elongation and decarboxylation pathways (Conrad and Jackson, 1971). Identification of the hydrocarbons from cockroach DGLP indicates the presence of three major compounds, *n*-pentacosane, 3-methyl-pentacosane, and 6,9-heptacosadiene (Table 7), which are also the major constituents of cuticular hydrocarbon in the American cockroach (Jackson and Baker, 1969). In addition, Baker *et al.* (1963) have reported hydrocarbons in the hemolymph of *P. americana*, and identified the same three compounds. It is now evident that the hydrocarbons in hemolymph are associated with the DGLP molecule.

4. Possible Multiple Role of DGLP—Comparison with Mammalian System

Diacylglycerol is transported in the hemolymph of many insects in association with a specific lipoprotein, DGLP, which serves as the primary vehicle of diacylglycerol transport. DGLPs isolated from different species also contain appreciable amounts of cholesterol, and it is possible that DGLP also functions to transport cholesterol. Insects, unlike mammals, are unable to synthesize sterols from acetate and are dependent upon a dietary source of sterols; therefore, the transport of newly absorbed sterol to sites of storage and utilization is of some importance.

Chino and Gilbert (1971) demonstrated that DGLP of *P. cynthia* can take up cholesterol from incubating midgut and, therefore, serve as a cholesterol carrier. Cockroach DGLP contains hydrocarbons with a composition similar to those found in the cuticle, and recently hydrocarbons have been detected in the DGLP of locusts (H. Chino, unpublished observations). Thus, it is likely that in many insects, DGLP serves also to transport hydrocarbons from the site of synthesis, presumably oenocytes (Diehl, 1975), to the site of deposition (cuticle).

The above observations suggest that in many insects, DGLP serves multiple functions in transporting such compounds as diacylglycerol, cholesterol, and hydrocarbon from sites of storage, absorption, or synthesis to sites where they are utilized as a metabolic fuel, precursors for triacylglycerol and phospholipid synthesis, or structural components of the cell membrane and cuticle. The proposed scheme of lipid transport in insects is illustrated in Fig. 4, and perusal of this scheme indicates that the term "lipid-carrying lipoprotein" may be a more appropriate designation for DGLP.

An appreciation of the physiological significance of insect DGLP may be gained by comparing the functions of DGLP with those reported for mammalian plasma lipoproteins. In mammals, several plasma lipoproteins, including chylomicrons, very low density lipoproteins, and low density lipoproteins, are involved in transporting such lipids as triacylglycerol and/or cholesterol ester from intestine or liver. These lipoproteins are formed in the cells of these tissues and are released into the lymph or blood. The triacylglycerol associated with these molecules is hydrolyzed by a lipoprotein lipase located in the capillary wall, and the resulting

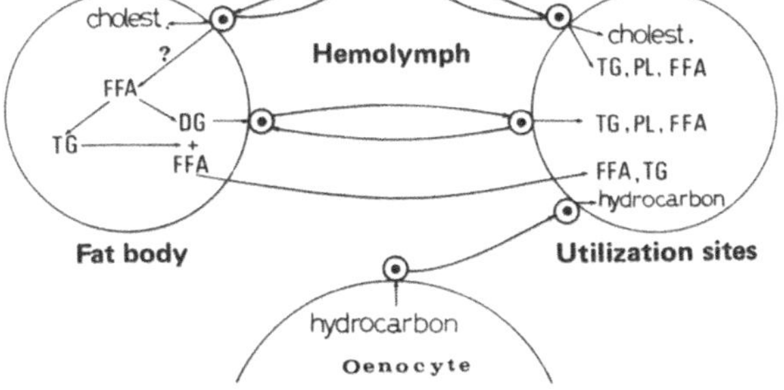

Fig. 4. Proposed scheme for lipid transport in insects. TG: triacylglycerol, DG: diacylglycerol, MG: monoacylglycerol, FFA: free fatty acid, PL: phospholipid.

Table 8. Lipid Composition of DGLP Isolated from Hemolymph and Eggs of *Philosamia* Silkworm[a]

	Relative amount	
Components	Hemolymph DGLP	Egg DGLP
Protein	100	100
Total lipids	11	78
Triacylglycerol	0.8	1.0
Diacylglycerol	1.6	45
Cholesterol	0.7	11
Phospholipids	7.8	21

[a] From Chino *et al.*, 1977.

remnant (mainly apoprotein and phospholipids) is broken into small fragments in the liver or extrahepatic tissues. By contrast, insect DGLP is formed in the fat body and is released into hemolymph, and it loads lipids, probably on the cell surface of different tissues, and unloads these lipids at the site of utilization; thus, the same molecule can perform this function repeatedly. This proposal is supported by studies with *P. cynthia*, in which DGLP isolated from mature eggs has been shown to contain less diacylglycerol than DGLP isolated from hemolymph; these observations suggest that DGLP unloads diacylglycerol in the ovary during oogenesis (Table 8), but the "empty" molecule retains its capacity to accept diacylglycerol from fat body (Chino *et al.*, 1977). Thus, from the viewpoint of physiological function, insect DGLP is a true carrier molecule, whereas mammalian plasma lipoprotein is not. It is apparent from the above discussion that the mechanism of lipid transport in insects is different from that of the mammal.

5. Possible Metabolic Regulation of DGLP

When [1-^{14}C]palmitic acid is administered to cockroaches, most of the radioactivity associated with DGLP is recovered from the diacylglycerol fraction, whether the label is administered by injection or by feeding (Table 1). This observation suggests a possible mechanism by which the supply, storage, and/or consumption of fatty acids may be regulated in polyphagous insects. When insects feed on fat-rich foods, DGLP serves to transport diacylglycerol as a metabolic fuel from the midgut to sites of utilization, such as the musculature; however, when insects are starved,

DGLP functions to transport diacylglycerol from fat body, the major site of triacylglycerol storage.

Chino and Gilbert (1965) demonstrated that in a physiologically inactive stage such as diapausing pupae, free fatty acid is the predominant form in which lipid is transported from fat body; but in an active stage, diacylglycerol is the major lipid released from this tissue (Table 9). On this basis, it was suggested that the release of diacylglycerol is an active process requiring ATP, whereas the release of free fatty acid is a passive process involving simple diffusion. Indeed, Chino and Gilbert (1965) showed, in *H. cecropia*, that the release of diacylglycerol from fat body is considerably inhibited by addition of respiratory poisons such as cyanide, whereas the release of free fatty acid is stimulated by such poisons (Table 9). Thus, the rate of release of diacylglycerol and free fatty acid from fat body is dependent on the rate of ATP-generation or the availability of metabolic energy in the tissue.

The release and transport of diacylglycerol from fat body of locusts is under the control of an adipokinetic hormone (Mayer and Candy, 1969; Stone *et al.*, 1976); a recent report suggests that one action of the hormone is to stimulate the formation of a specific hemolymph lipoprotein that is capable of accepting the elevated levels of diacylglycerol (Van der Horst *et al.*, 1979). An *in vitro* system using purified DGLP may be of considerable benefit in elucidating the mode of action of adipokinetic hormone.

These observations lead to the conclusion that in many insects DGLP (more correctly called lipid-carrying lipoprotein) is involved in the reg-

Table 9. Release of Diacylglycerol and Free Fatty Acid from Prelabeled Fat Body of *Cecropia* Silkworm and the Effect of Respiratory Inhibitors[a,b]

		Diacylglycerol released		Free fatty acid released	
Prelabeled fat body	Inhibitor	dpm	% release or relative value	dpm	% release or relative value
Diapausing pupal fat body	—	1,266	4.5%	10,020	27.0%
Adult fat body	—	106,570	29.8%	3,160	5.9%
Adult fat body	None	46,280	100	2,840	100
	5 mM KCN	15,510	33	7,330	257
	5 mM NaN_3	19,190	41	4,560	160

[a] The prelabeled fat bodies were incubated with 0.3 ml pupal hemolymph for 60 min. % release represents the percentage of the radioactivities of the lipids released into the incubation media relative to those originally contained in the prelabeled fat bodies.
[b] From Chino and Gilbert, 1965.

ulation of energy metabolism according to particular nutritional, physiological, and hormonal conditions.

References

Baker, G.L., Vroman, H.E., and Padmore, J., 1963, Hydrocarbons of the American cockroach, *Biochem. Biophys. Res. Commun.* **13**:360.

Chino, H., and Downer, R.G.H., 1979, The role of diacylglycerol in absorption of dietary glyceride in the american cockroach, *Insect Biochem.* **9**:379.

Chino, H., and Gilbert, L.I., 1964, Diglyceride release from insect fat body, *Science*, **143**:359.

Chino, H., and Gilbert, L.I., 1965, Lipid release and transport in insects, *Biochim. Biophys. Acta* **98**:94.

Chino, H., and Gilbert, L.I., 1971, The uptake and transport of cholesterol by hemolymph lipoproteins, *Insect Biochem.* **1**:337.

Chino, H., Murakami, S., and Harashima, K., 1969, Diglyceride-carrying-lipoproteins in insect hemolymph: Isolation, purification and properties, *Biochim. Biophys. Acta* **176**:1.

Chino, H., Yamagata, M., and Takahashi, K., 1976, Isolation and characterization of insect vitellogenin: Its identity with hemolymph lipoprotein II, *Biochim. Biophys. Acta* **441**:349.

Chino, H., Downer, R.G.H., and Takahashi, K., 1977, The role of diacylglycerol-carrying-lipoprotein in lipid transport during insect vitellogenesis, *Biochim. Biophys. Acta* **487**:508.

Chino, H., Katase, H., Downer, R.G.H., and Takahashi, K., 1981, Diacylglycerol-carrying-lipoprotein of hemolymph of american cockroach: Purification, characterization and function, *J. Lipid Res.* **22**:7.

Chinzei, Y., Chino, H., and Wyatt, G.R., 1981, Purification and properties of vitellogenin and vetellin from locusts, *Insect Biochem.*, **11**:1.

Conrad, C.W., and Jackson, L.L., 1971, Hydrocarbon biosynthesis in *Periplaneta americana, J. Insect Physiol.* **17**:1907.

Diehl, P.A., 1975, Synthesis and release of hydrocarbons by the oenocytes of the desert locust, *J. Insect Physiol.* **21**:1237.

Gellissen, G., and Emmerich, H., 1980, Purification and properties of a diglyceride-binding-lipoprotein (LP-I) of the hemolymph of adult male locusta, *J. Comp. Physiol.* **136**:1.

Hoffman, A.G.D., and Downer, R.G.H., 1979, Synthesis of diacylglycerols by monoacylglycerol acyl transferase in midgut and fat body tissues of the american cockroach, *Insect Biochem.* **9**:129.

Jackson, L.L., and Baker, G.L., 1969, Cuticular lipids in insects, *Lipids* **5**:239.

Mayer, R.J., and Candy, D.J., 1969, Control of hemolymph lipid concentration during locust flight: An adipokinetic hormone from the corpora cardiaca, *J. Insect Physiol.* **15**:611.

Mwangi, R.W., and Goldsworthy, G.J., 1977, Diglyceride-transporting-lipoprotein in *Locusta, J. Comp. Physiol.* **114**:177.

Peled, Y., and Tietz, A., 1973, Fat transport in the locust: The role of protein synthesis, *Biochim. Biophys. Acta* **296**:499.

Peled, Y., and Tietz, A., 1975, Isolation and properties of a lipoprotein from the hemolymph of the locust, *Insect Biochem.* **5**:61.

Stone, J.V., Mordue, W., Batley, K.E., and Morris, H.R., 1976, Structure of locust adipokinetic hormone, a neurohormone that regulates lipid utilization during flight, *Nature (London)* **263**:207.

Thomas, K.K., and Gilbert, L.I., 1968, Isolation and characterization of the hemolymph lipoproteins of american silkmoth, *Hyalophora cecropia*, *Arch. Biochem. Biophys.* **127**:512.
Tietz, A., 1967, Fat transport in the locust: The role of diglycerides, *Eur. J. Biochem.* **2**:236.
Van Der Horst, D.J., Van Doorn, J.M., and Beenakkers, A.M.T., 1979, Effects of the adipokinetic hormone on the release and turnover of hemolymph diglycerides and on the formation of the diglyceride-transporting lipoprotein system during locust flight, *Insect Biochem.* **9**:627.

7

Energy Metabolism in the Insect Nervous System

R.H.C. Strang

1. Introduction

The study of the fundamental metabolism of the insect nervous system represents a small island, well off the normal shipping lanes, occasionally noticed by navigators passing between the larger continental masses of metabolic studies. Its lonely beaches are not at much risk from the tourist trade. Nevertheless, as any biologist knows, small islands offer interesting adaptations of better explored systems, suited to their peculiar needs. For that reason, the prevailing tone of the following chapter will be one of comparison. Such features of the energy metabolism of the insect nervous system as are well known will be compared to those of two other tissues much better charted: the mammalian central nervous system, and the insect flight muscle. The energy metabolism of both of these may represent an evolutionary climax in terms of their biochemical and physiological development, and their importance to the life of the animals. In the case of the latter, the comparison with the metabolism of the insect nervous system is of more than academic interest, as the activity of the flight muscles, with their vast consumption of energy, will have important implications for the functioning of the nervous tissue housed within the same body.

I hope that the comparison will place the metabolism of the insect nervous system in some sort of context. I doubt if such studies have much point in complete isolation. After all, to a Pygmy a Bushman is a giant,

R.H.C. Strang • Department of Biochemistry, University of Glasgow, Glasgow, Scotland.

until he has seen a Watusi. Two caveats must be uttered at the outset. One of the most obvious reasons for the lack of exploration of the biochemistry of the insect nervous system is the difficulty of obtaining sufficient tissue. An individual insect will only yield a few milligrams of tissue after much finicking dissection. For that reason, virtually the only tissues examined in detail are those of large insects, such as the locust and the cockroach. It is highly unlikely that any of the conclusions in this chapter can be taken as generalizations true for the nerves of all insects. Much more elaborate studies on other aspects of insect biochemistry, such as flight muscle metabolism, have shown that the diversity of insect biochemistry defies all such attempts at generalization.

Furthermore, even in these two types of insect the picture is far from complete. Therefore, the gazetteer which follows has more the vagueness of a Ptolemy than the completeness of a Baedeker.

2. Oxygen Uptake Studies

As the insect has no vascular system comparable to that of the mammal, there is no possibility of directly determining the oxygen uptake of the nervous tissue *in vivo* by arterial–venous difference studies, as has been done in the mammal. In Table 1 are listed the rates of *in vitro* oxygen uptake for the nervous tissues of the locust, cockroach, and other animals under a variety of conditions. At first glance, there seems to be a wide disparity between the different figures obtained for the insect tissues by different workers. This is hardly surprising in the determination of a parameter such as oxygen uptake *in vitro*, which is affected by such a number of factors. From Figs. 1 and 2, it is clear that oxygen uptake of the isolated ganglia of the locust nervous system is dependent upon the temperature and the concentration of the gas in the medium. The Q_{10} between 25° and 35°C is three rather than two; moreover, the rate has been found to vary with the length of time of incubation, falling off after an hour, and also with the composition of the saline (Clement, 1979). There may also be methodological reasons for the different estimates. Clement and Strang (1978) and Steele and Chan (1980) used an oxygen electrode, which is ideal for obtaining initial rates of consumption of oxygen. Bradford *et al.* (1969), on the other hand, made their estimate by manometry over a three hour period. In view of the decline in oxygen uptake with time, their lower figure is hardly surprising. It cannot be expected that oxygen uptake studies will produce the sort of hard and fast figures that might result, for instance, from the measurement of the maximal specific activity of an enzyme. This is especially true of nervous tissue, in which so much of the oxygen uptake depends on the degree of electrical activity of the tissue.

Table 1. Rates of Oxygen Uptake by Various Preparations of Nervous and Flight Muscle Tissue[a,b]

Tissue	Temperature (°C)	Notes on conditions	Stimulation	O_2 consumption (μmol/g/hr)	References[c]
Locust thoracic ganglia	37	Initial rate	−	270	1
Locust thoracic ganglia	37	Rate after 60 min	−	142	2
Locust thoracic ganglia	25	Rate averaged over 3 hr	−	35	3
Cockroach nerve cord	30	200–300 μM O_2	−	100	4
			+ (Octopamine)	160	
Rat cerebral cortex slices	37		−	96	1
Rat cerebral cortex slices	37		−	89	3
Rat retina	37		−	125	5
Guinea pig cerebral cortex slices	37		−	61	6
			+ (electrical)	124	
Rat sympathetic ganglion	35–37		−	27	7
			+ (electrical)	45	
Snail (*H. pomatia*) central ganglia	25		−	12	3
Fish brain	16–18		−	9	8
			+ (electrical)	15	
Locust flight muscle	25[d]	*in vivo*	+ (flying)	5040	9
Cockroach flight muscle	25[d]	*in vivo*	+ (flying)	5400	9
Locust flight muscle	35		+	540[e]	10

[a] Unless otherwise stated, all the estimates were carried out *in vitro*, in the appropriate saline, under optimal concentrations of glucose and oxygen.

[b] The locusts used were *Schistocerca americana gregaria*, and the cockroaches were *Periplaneta americana*.

[c] References: (1) Clement and Strang, 1978; (2) Clement, 1979; (3) Bradford *et al.*, 1969; (4) Steele and Chan, 1980; (5) Elliott, 1955; (6) McIlwain, 1953; (7) Larrabee, 1958; (8) McIlwain and Bachelard, 1971a; (9) Crabtree and Newsholme, 1972; (10) Candy, 1978.

[d] This represents the air temperature. The temperature in the thorax may have been higher.

[e] Oxygen uptake estimated on the basis of the complete oxidation of the glucose consumed.

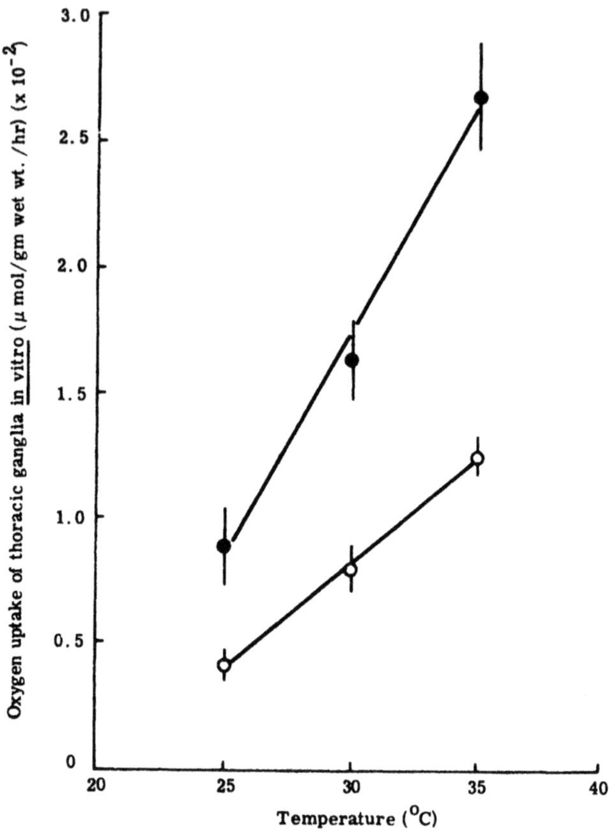

Fig. 1. Relationship of O_2 uptake to temperature for isolated unstimulated thoracic ganglia of the locust *Schistocerca americana gregaria*. The medium was preequilibrated with air (○), or with 100% O_2 (●). Vertical lines represent ± standard deviation. (Figure reproduced from Clement and Strang, 1978.)

More pertinent questions might be: how does the rate recorded *in vitro* compare with the possible rates *in vivo*, and how does the insect tissue compare to the mammalian under the same circumstances?

When insect tissues are removed from the body, the tracheal system is disrupted, and oxygen must diffuse into the tissues over longer distances (up to 1 mm) than is the case in the living animal. For this reason, when the oxygen in the medium is in equilibrium with the atmosphere, its concentration is one of the main factors limiting the rate of oxygen uptake (Clement and Strang, 1978) However, when the ambient concentration of dissolved oxygen exceeds 700 μM, a concentration only obtainable by bubbling pure oxygen through the medium, it is no longer rate limiting to the utilization of the gas by intact ganglia. Unlike the mammalian nervous system, that of the insect *in vitro* displays a high level of spontaneous electrical activity, which persists for different periods of time depending upon the constituents of the artificial medium, and which has been shown to be equivalent to between 30% and 50% of that found in the unstimulated and (almost) intact insect (Clement and Strang, 1978). In addition, other criteria of the vitality of the tissue *in vitro* under optimal conditions, such as the concentrations of high energy phosphates, imply that in many respects, the condition of the tissue approximates closely to that found *in vivo*.

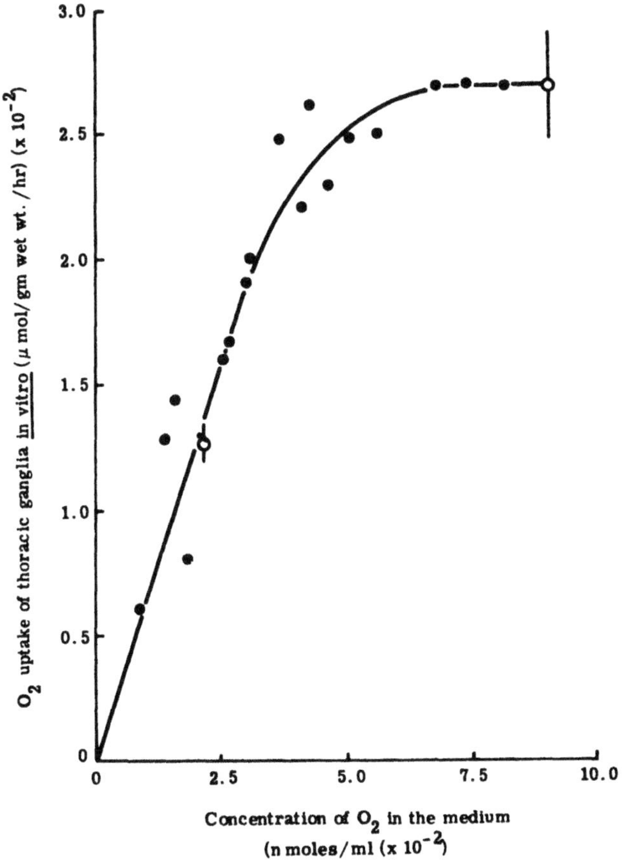

Fig. 2. Relationship of O_2 uptake to O_2 concentration in the medium for isolated thoracic ganglia of the locust *Schistocerca americana gregaria*. Measurements were made at 35°C in 10 mM glucose in saline. Filled symbols (●) denote the O_2 uptake at regular intervals during one experiment. Superimposed upon the curve are the averages (± standard deviation) of many measured rates made with the medium saturated with atmospheric, and 100% O_2 (○). (Figure reproduced from Clement and Strang, 1978.)

How do the various properties of the mammalian nervous tissues compare to this? Unlike the preparations from many invertebrates, those from mammals do not have a high spontaneous activity *in vitro*. In various examples in Table 1, electrical stimulation doubles the rate of oxygen consumption to a figure that is close to that found in the intact animal. It is only with this stimulation that slices of mammalian cerebral cortex approach the rates of uptake found for unstimulated locust and cockroach ganglia under optimal conditions.

Summing up, the results so far indicate that given the same temperature, the oxygen consumption of the nervous system of the insect *in situ* is at least equal to that of the mammal, and may well be much greater. It is certainly true that *in vitro* greater rates of oxygen uptake have been recorded for insect ganglia than for various preparations of mammalian nervous tissue under similar conditions.

One may object that some of the measurements of metabolic rate in the locust thoracic ganglia were made at 37°C, and that this is a temperature unlikely to be reached in a poikilothermic animal. During flight, however, temperatures within the thorax may exceed the temperature of the air by up to 10°C. Thus, temperatures of about 40°C are not impossible in the flying locust (Church, 1960).

3. Activities of Enzymes Associated with Glycolysis and the Citric Acid Cycle

Great care must obviously be taken in any attempt at extrapolation from the maximum specific activities of enzymes assayed *in vitro* to the metabolism in the living animal, in the absence of the most complete information about all the factors controlling those activities. Nevertheless, such maximal specific activities of rate-limiting enzymes may be used to mark the outermost marches of a particular area of metabolism, and thus give it some shape, even though the normal commerce of the tissue is carried out well within these boundaries.

In Tables 2 and 3 are collected the maximal specific activities of crucial enzymes associated with energy metabolism from the various tissues under comparison.

In the case of the citric acid cycle, the enzymes thought to control the rate of metabolism are citrate synthetase and isocitrate dehydrogenase. Studies on the latter are complicated by the fact that it is present in both cytoplasm and mitochondria, and has, in addition, two forms that accept as cofactor either NAD or NADP. The conventional wisdom has been that only the NAD-linked enzyme is concerned with the energy generating aspect of the cycle (e.g., Lehninger, 1975; Banks *et al.*, 1968), but oxygen uptake studies make it clear that in some tissues, the activity

Table 2. Maximal Specific Activities of Citrate Synthetase and Isocitrate Dehydrogenase in Nervous and Flight Muscle Tissues[a]

Enzyme	Tissue	Temperature (°C)	Notes		Specific activity (μmol/g/min)	References[b]
Citrate synthetase	Locust cerebral ganglia	25			21	1
	Cockroach[c] cerebral ganglia	25			24	1
	Mouse brain	25			24	1
	Rat brain	25			14	1
	Locust flight muscle	25			242	2
	Cockroach flight muscle	25			185	2
Isocitrate dehydrogenase	Locust cerebral ganglia	25	NAD	total	6	1
			NADP		7	
	Cockroach[c] cerebral ganglia	25	NAD	total	–	1
			NADP		3	
	Locust thoracic ganglia	37	NAD	total	5	3
			NADP		11	
			NAD	mitochondrial	3	
			NADP		6	
	Mouse brain	25	NAD	total	3	1
			NADP		2	
	Rat brain	25	NAD	total	1	1
			NADP		1	
	Locust flight muscle	25	NAD	total	54	2
			NADP		4	
	Cockroach flight muscle	25	NAD	total	74	2
			NADP		3	
	Locust flight muscle	37	NAD	total	28	3
			NADP		3	

[a] All assays, unless otherwise stated, were made with crude homogenates of the tissue concerned. Except where indicated, the locusts used were *Schistocerca americana gregaria* and the cockroaches were *Periplaneta americana*.
[b] References: (1) Sugden and Newsholme, 1975; (2) Alp, 1973; (3) Strang *et al.*, 1979.
[c] *Blaberus discoidalis*.

Table 3. Maximal Activities of Glycolytic Enzymes from Various Nervous and Flight Muscle Tissues[a]

Enzyme	Tissue	Temperature (°C)	Notes	Specific activity (μmol/g/min)	References[b]
Hexokinase	Locust thoracic ganglia	37	2.5 and 5.0 mM ATP	1.5	1
			ATP regenerating system	3.0	
	Locust cerebral ganglia	25	ATP regenerating system	5.6	2
	Cockroach	25	ATP regenerating system	5.1	2
	Mouse brain	25	Whole brain	15.0	2
	Rat brain	25	Whole brain	14.0	2
	Guinea pig brain	37	Cortex	21–26	3
	Mouse brain	37	Whole brain	27.0	4
	Locust flight muscle	37	2.5 mM ATP	6.5	1
			ATP regenerating system	16.0	
	Locust flight muscle	25	ATP regenerating system	11.5	5
Phosphofructokinase	Locust thoracic ganglia	37		7.0	1
	Locust cerebral ganglia	25		4.6	2
	Cockroach cerebral ganglia	25		6.0	2
	Mouse brain	25		15.0	2
	Rat brain	25		18.0	
	Locust flight muscle	37		20–24	1
	Locust flight muscle	25		13.0	2
Lactate dehydrogenase	Locust thoracic ganglia	37		10.4	1
	Mammalian brain	37		90–180	4, 6, 7
	Locust flight muscle	25		3	5
	Locust flight muscle	37		0	1
Glycerol phosphate dehydrogenase	Locust thoracic ganglia	37		11	1
			Total	140	
	Locust flight muscle	25	Mitochondrial	43	5
	Locust flight muscle	37	Total	108	1

[a] Except where indicated, the assays were made with crude tissue homogenates. The locusts used were *Schistocerca americana gregaria*, and the cockroaches *Blaberus discoidalis*.

[b] References: (1) Strang, *et al.*, 1979; (2) Sugden and Newsholme, 1973; (3) Bachelard, 1967; (4) Lowry *et al.*, 1964; (5) Crabtree and Newsholme, 1972; (6) Johnson, 1960; (7) Balázs *et al.*, 1968.

of the NAD-linked enzyme is inadequate to account for the rates found (Newsholme and Start, 1974). Indeed, as these authors have shown, in some fish muscles there is virtually no activity of this particular form of the enzyme at all. In the absence of better information, it is probably reasonable to add together the activities of the two forms. The fact that it is only the mitochondrial activity which is directly relevant to the citric acid cycle, and thus to oxidative metabolism, has been taken less into consideration. Strang *et al.* (1979) found a more or less equal distribution of the activity between the cytoplasm and mitochondria of the locust nervous system.

In all the quoted cases, the activity of the citrate synthetase is higher than that of the total activity of the isocitrate dehydrogenase, making it likely that it is the latter which is the rate-limiting enzyme, although other factors, such as the supply of substrate and allosteric control mechanisms, might be of importance within the cell. The fact that in those cases in which it has been measured, the concentration of citrate in the insect tissues and hemolymph is much higher than in the mammal, reinforces the conclusion that it is the isocitrate dehydrogenase and not citrate synthetase which limits the activity of the cycle (Florkin and Jeuniaux, 1974; E.A. Newsholme, personal communication, 1979: Rowan and Newsholme, 1979; Worm and Beenakkers, 1980). If a Q_{10} of between one and two is assumed, there is greater agreement among different authors for the activity of the NADP-linked form of the dehydrogenase than for the NAD-linked form. Both available estimates (Sugden and Newsholme, 1975: Strang *et al.*, 1979), however, agree that the total activity of the NADP-linked form is greater than the NAD-linked, in the locust nervous system. In this it is quite different from the flight muscle, in which not only is the full activity of the enzyme much greater than in the nervous tissue, but it is the NAD-linked form that is predominant.

Although the activity of the enzyme in the locust nervous system is much lower than that in the flight muscle, it is greater than the activities found in nervous tissues from other animals, both vertebrate and invertebrate (Sugden and Newsholme, 1975). This is consistent with the low rates of oxygen uptake found for different invertebrate nervous systems (Bradford *et al.*, 1969). If the activity of isocitrate dehydrogenase does impose a limitation upon aerobic metabolism, then the total mitochondrial activity of the enzyme is 9 μmol/g/min at 37°C, which would require an oxygen consumption of 18 μmol/g/min, far in excess of even the highest rate of 4.5 μmol/g/min recorded *in vitro* (Clement and Strang, 1978).

Comparing the maximal activities of the key citric acid cycle enzymes with those of glycolysis in the different nervous tissues, a sharp contrast is apparent between the mammal and the insects in question. This is especially marked in the case of hexokinase. Tissues from the brains of many mammals have consistently shown a total hexokinase activity of between 20 and 30 μmol/g/min at 37°C, far in excess of the normal requirement *in vivo* (McIlwain and Bachelard, 1971b). The activity of the enzyme in the ganglia of the locust is much lower, although there is disagreement as to the figure. The activity of 5–6 μmol/g/min at 25°C for the cerebral ganglia is obviously much higher than that of 1.5–3. O μmol/g/min at 37°C, if the normal temperature coefficient applies.

In the case of hexokinase, whose activity is subject to control by a number of ligands, it is difficult to be sure of the relationship of the activity *in vitro* to that *in vivo*. Clearly, the presence of an ATP-regenerating system increases the activity (Strang *et al.*, 1979), and this may be as much due to the low concentration of ADP, which inhibits the activity (Sols and Crane, 1954), as to the maintenance of a high concentration of ATP. In addition, recent work in the author's laboratory has shown that in common with hexokinase from other sources, that from the thoracic ganglia of the locust is inhibited by low concentrations of glucose-6-phosphate (Flanigan, T., and Strang, R.H.C. unpublished observation). If the lower activities of the locust enzyme do in fact represent the activity *in vivo*, even under extreme circumstances, then that activity is only about twice the flux of hexose units *in vitro* (calculated from the rate of oxygen uptake), a remarkably small margin in comparison to that in the mammalian brain, in which the ratio is about 50.

This, however, may be too simplistic an interpretation of inadequate data, especially in view of the fact that, unlike the situation in the locust nervous system, in which all the hexokinase activity is in the cytoplasm (Strang *et al.*, 1979), the enzyme in the mammalian brain is divided between the cytoplasm and the mitochondria (Bachelard, 1976). It has been suggested (Bachelard, 1976) that only the cytoplasmic enzyme is normally active in glycolysis. This would make the contrast between the mammal and the insect much less sharp, as the maximal activity in the mammalian cytoplasm is only 3–4 μmol/g/min at 37°C. Nevertheless, on balance, it seems that the thoracic ganglia of the locust have a much lower total capacity to phosphorylate circulating glucose than does the mammalian brain.

The activity of phosphofructokinase is a better indicator of the full glycolytic potential, as it encompasses the input from glycogenolysis. In the ganglia of the locust and of the cockroach, the estimates obtained are in fair agreement and represent a potential equal to or greater than that of hexokinase.

In order to obtain a comparison of anaerobic to aerobic carbohydrate metabolism, Sugden and Newsholme (1975) multiplied the maximal activity of phosphofructokinase by two, thus expressing the potential of the glycolytic pathway in terms of triose units formed. The figure for the locust ganglia implies a full glycolytic capacity of 14–18 μmol pyruvate formed/g/min at 37°C. This compares with a total mitochondrial isocitrate dehydrogenase activity of about 9 μmol/g/min at the same temperature. Thus, the total possible flux through the glycolytic pathway is only twice the capacity of the citric acid cycle. This contrasts with the situation in the mammalian brain, in which the ratio of activities of the two enzymes is much higher: six, in the case of the mouse, and 18 in the case of the rat (Sugden and Newsholme, 1975). The implications are that while both the mammalian and locust central nervous systems are normally highly aerobic tissues, the former has a huge anaerobic capacity, which potentially could far outstrip the aerobic; but in the latter, there is a much closer match between the anaerobic and aerobic parts of carbohydrate metabolism.

This general conclusion receives reinforcement from a consideration of the maximal specific activities of lactate dehydrogenase in the different tissues. In the mammal, the estimates of cortical lactate dehydrogenase range from 90 to 180 μmol/g/min at 37°C, while that for the thoracic ganglia of the locust is only about 10 μmol/g/min. This lower capacity for lactate production predicted from the specific activity of the enzyme is confirmed by the behavior *in vitro* of various preparations of the two sorts of nervous tissue. In mammalian tissue, between 30% and 60% of the total glucose consumed forms lactate (McIlwain *et al.*, 1952; Horowicz and Larabee, 1957; Härkönen *et al.*, 1969), while in the case of the locust ganglia, the relevant figure is 3% to 5% (Strang *et al.*, 1979). (It is probable, however, that lactate is not the main product of anaerobic metabolism; see Section 5.1.)

In contrast to the nervous system, the flight muscle of the locust and other insects has long been known as an almost obligately aerobic tissue, demonstrating rates of oxygen consumption during flight unequalled by any other known tissue, but under tight control at rest (Davis and Fraenkel, 1940; Weis-Fogh, 1952). In keeping with this picture, lactate dehydrogenase activity is almost totally absent from the tissue. As a corollary, the tissue has a very high activity of glycerol phosphate dehydrogenase, the key enzyme of the glycerol phosphate shuttle, which in the flight muscle is vitally important in transferring reducing equivalents from cytoplasmic NADH into the mitochondria, and thus ensuring the maintenance of the highest possible glycolytic flux.

In the nervous system the picture is much less clear. Although the mammalian brain possesses glycerol phosphate dehydrogenase activity, histochemical evidence has recently indicated that it may lie solely within oligodendroglial cells (Hirsch *et al.*, 1980). If this is so, then the enzyme must be relegated to a minor role in the energy metabolism of the brain. The activity of the enzyme in the locust ganglia was only 10% of that found in the muscle tissue, which, in turn, was much lower than figures found by others for the same or comparable flight muscles (Crabtree and Newsholme, 1972). In the nervous tissue it is equalled by the activity of the lactate dehydrogenase. It could be that the compartmentation found in the mammalian central nervous system applies also to the insect, which, although simpler than the mammal in that it has fewer nerve cells, is still heterogeneous regarding cell types. It remains an open question whether or not the glycerol phosphate shuttle in the flight muscle has any counterpart in the nervous tissue.

4. Fuels

4.1. Carbohydrate

The maximal activity of key enzymes *in vitro* can only delineate metabolism of a tissue with the broadest of strokes. Such figures represent potentialities rather than actualities. Determination of the latter depends on the conditions in the intact tissue and surrounding medium. On the basis of the results already presented, the nervous tissues of the locust and cockroach show the capacity for a vigorous aerobic metabolism consistent with the life of such active invertebrates. What fuel supplies the energy required? By analogy with the mammal, and from a knowledge of its total concentration in the hemolymph, common sense suggests that carbohydrate must be a prime candidate. The mammalian brain is a tissue with a large and continuous consumption of circulating glucose. The figure often quoted in this context is that with only 4% of the body mass it is responsible for 25% of the glucose utilization of the whole human body at rest (Bachelard, 1978). In the short term, at least, the mammalian brain is very dependent upon an adequate supply of glucose in the bloodstream, and overt behavioral symptoms appear should that concentration fall abruptly.

Although glucose is present in the hemolymph of the insects under discussion in concentrations ranging from 0–10 mM (Howden and Kilby, 1956; Treherne, 1958; Strang and Clement 1980), and occasionally higher (Hansen, 1964; Mayer and Candy, 1969), the major sugar in the hemolymph of most fed, resting insects is trehalose, the disaccharide of glucose whose presence seems often to be a characteristic of vigorously flying insects (but not invariably so) (Bedford, 1977). This carbohydrate occurs in remarkably high concentrations, ranging, according to many reports, from 40 to 80 mM and higher. (Howden and Kilby, 1956; Treherne, 1960; Hansen, 1964; Mayer and Candy, 1969; Jutsum and Goldsworthy, 1976; Strang and Clement, 1980). It would seem to be a reasonable assumption that in such concentrations, trehalose would contribute largely to the fuel consumed by the insect nervous system, as has been found to be the case for cockroaches and locusts by Treherne (1960) and Strang and Clement, (1980).

When the insect starts to fly, however, there is a steep fall in the concentration of trehalose. The extent of this fall is in some dispute, ranging from a reported total disappearance of all carbohydrate on extended flight (Weis-Fogh, 1964), to one of only 50% of the resting concentration (Jutsum and Goldsworthy, 1976; Robinson and Goldsworthy, 1976). Under the same conditions, the fall in glucose concentration is neither so immediate nor so complete (Mayer and Candy, 1969; Strang and Clement, 1980), and at the start of flying, there is an increase in the glucose concentration (Poláček and Kubišta, 1960; Mayer and Candy, 1969). These findings raise two points concerning the carbohydrate as an energy substrate for the central nervous system: the first is the relative importance of trehalose and glucose to the nervous system, under different conditions; and the second is the question as to whether or not the concentration of carbohydrate always constitutes an adequate reservoir of substrate for the functioning of the nervous tissue. Tables 4 and 5 represent an attempt to answer these two questions for isolated ganglia of the locust nervous system. The tissues were incubated in the presence

Table 4. Rates of O_2 Consumption by Locust Thoracic Ganglia *in Vitro*[a,b]

Equivalent *in vivo* conditions	Concentrations of carbohydrate in the medium (mM)		Rate of O_2 uptake (μmol/g/hr)
	Glucose	Trehalose	
—	10	—	268
—	—	—	103**
Resting	4	50	250
	4	—	220*
	—	50	184*
1 hr of flight	3	25	258
	3	—	200*
	—	25	140**
2 hr of flight	1	15	164**
	1	—	165**
	—	15	146**

[a] Thoracic ganglia of *Schistocerca gregaria* were incubated in insect saline in an oxygen electrode containing concentrations of glucose and trehalose that were the averages of those found in the hemolymph of locust under the conditions noted. The temperature of incubation was 37°C. Those rates marked by asterisk were significantly different from the maximal rate: $^{*}p < 0.01$, $^{**}p < 0.001$.

[b] Data taken from Strang and Clement, 1980.

Table 5. Proportions of CO_2 Derived from Glucose and Trehalose by Intact Ganglia *in Vitro*[a]

Concentration of carbohydrate in the medium (mM)			Ratio of glucose to trehalose	
Glucose	Trehalose	% CO_2 derived from glucose	(as glucose) (%)	Equivalent conditions *in vivo*
4	60	18	3	Resting[b]
3	25	25	6	After 1 hr flying[b]
15	40	68	19	resting[c]
20	15	95	67	After 30 min flying[c]

[a] The proportional utilization of the sugars was determined by incubating the ganglia in the presence of [^{14}C]glucose at constant specific activity and unlabeled trehalose, and trapping the CO_2 formed.
[b] Strang and Clement, 1980.
[c] Mayer and Candy, 1969.

of concentrations of glucose and trehalose that represented average concentrations found in the hemolymph of resting and flying insects. The results for the oxygen uptake show that while trehalose is always higher in concentration (at the time intervals chosen) than glucose, even at its resting concentration of more than 50 mM it was less effective at supporting oxygen uptake than was the much lower concentration of glucose. Furthermore, although a combination of glucose and trehalose in concentrations found after an hour of flying were capable of supporting the maximal rate of oxygen uptake *in vitro*, this was no longer true with the concentrations present after a second hour of flight. Still lower concentrations of carbohydrate have been found after longer periods of enforced flight. (Weis-Fogh, 1964; Tietz, 1967). In this result lies the implication (and it is no more than that, as yet) that occasionally the concentration of circulating carbohydrate falls too low to be an adequate single substrate for the nervous system.

With regard to the relative importance of trehalose and glucose to the nervous system, the results in Table 5 generally corroborate the results of oxygen uptake. Only when the concentration of trehalose is high relative to glucose does the trehalose contribute significantly to the carbon dioxide produced. The same probably applies to the cockroach nervous system, for which the relative uptake of trehalose and glucose under resting conditions is almost identical to that in the locust (Treherne, 1960). Under some conditions reported in the flying locust (Mayer and Candy, 1969), trehalose is hardly used at all by the thoracic ganglia. Conversely, glucose, despite its low concentration relative to trehalose in the hemolymph, is taken up and utilized much more efficiently by the ganglia. In light of this, Treherne (1960), suggested that there might be some uptake mechanism for glucose, similar to that which is present in the mammal, to allow its rapid entry into the insect nervous system. While this is possible, on the basis of evidence which will be discussed later, such a mechanism, if present, operates much less effectively than does the facilitated uptake in the mammalian brain.

The results of other authors contradict the suggestion that the decline in carbohydrate concentration in the hemolymph owing to flight really poses a threat to the well-being of the nervous system. Robinson and Goldsworthy, (1976, 1977a,b) reported that the trehalose concentration did not fall below 40 mM, and that in the locust, the adipokinetic hormone stimulated the release of diacylglycerol and its utilization by the muscle. Thus, the concentration of circulating carbohydrate is spared, providing a protective device for other tissues, such as the nervous system. In view of the great difference between species, and between individuals of the same species in this regard (A.M.T. Beenakkers, personal communication), it is unwise to be too dogmatic; but there is no doubt that the sparing mechanism is rather ineffective, as individual insects have been found with very low concentrations of circulating carbohydrate after prolonged flying (Strang and Clement, 1980). Even if a sparing or homeostatic mechanism operates during flight in a locust, the same does not seem to be true during starvation. A five day fast in locusts has been reported to produce falls of 90% in the concentration of circulating carboyhdrate (Mwangi and Goldsworthy, 1977). Under these circumstances, it seems inevitable that some other substrate must be used by the nervous tissue.

Despite the variation in the results, it is likely that the nervous systems of both locust and cockroach have to cope with wide fluctuations in concentration of available carboyhdrate under extreme but conceivable circumstances. This is quite different from the situation in the mammal, in which a variety of homeostatic mechanisms function to maintain the concentration of glucose in the bloodstream within narrow limits. In man, running to exhaustion produces only a 6% to 13% fall in glucose concentration (Edwards *et al.*, 1934; Hermansen *et al.*, 1967), and four days of starvation resulted in only a 30% fall in the circulating carbohydrate in the rat bloodstream (Hawkins *et al.*, 1971).

Whatever glucose-sparing mechanisms are present in the locust and other insects, which are analogous to the glucose–fatty acid cycle in mammals, the mechanism must be quite different. In mammalian muscle and nervous system, increased concentrations of citrate resulting from fatty acid utilization inhibit the activity of phosphofructokinase. No such effect is present in insect tissues (Walker and Bailey, 1969; Newsholme *et al.*, 1977; Strang *et al.*,1979). In this context, it is interesting to notice that the concentrations of citrate normally present in locust and other insect tissues are those that would cause inhibition in mammalian tissue.

4.2. Enzymes of Carbohydrate Utilization

Searching further into the reason for the relatively poor capacity of the insect nervous system to make effective use of quite high concentrations of carbohydrate, the most obvious places to look are the uptake of the carbohydrate into the tissue and the nature of the enzymes that then utilize it.

In the mammalian brain, there is a specific mechanism of facilitated diffusion for glucose and a few analogues (Fishman, 1964; Crone, 1965). The characteristics of the system are such that the supply of glucose only becomes rate limiting when the concentration of glucose falls below 2 mM (Bachelard, 1976). At a concentration below this figure, overt symptoms begin to appear in the living animal if the fall has been an acute one. This figure also represents the minimum concentration that will support the maximum rate of oxygen uptake by isolated slices of guinea pig cortex (McIlwain, 1953). When the thoracic ganglia of the locust are incubated *in vitro*, a concentration of 10 mM glucose is required to support the maximum rate of oxygen uptake. With respect to the supply of nutrient from the surrounding medium, the ganglion suspended in artificial medium is identical to the tissue *in vivo*. In both cases, the substrates must diffuse into the tissue from the outside. The concentration of 10 mM glucose is remarkably high in comparison to mammalian tissue, and is one seldom found in the hemolymph of the insects. The high figure makes it unlikely that if there is a specific uptake mechanism for glucose, it is very effective. Experiments with 2-deoxy-glucose could be used to determine the existence of such a mechanism.

Further utilization of glucose, whether derived from the free sugar or from the hydrolysis of trehalose, depends on phosphorylation by hexokinase. As has been already discussed, its activity in the locust nervous system is lower than that in the nervous systems of a wide range of animals, invertebrate and vertebrate (Sugden and Newsholme, 1973). The K_m of the locust enzyme for glucose is 2.2×10^{-4} M, which is higher than those in mammalian brains, which range from 2.0×10^{-5} M to 7.4×10^{-5} M (Table 6). In the absence of further information about the intracellular concentrations of glucose, it is impossible to say if this lower affinity of the locust enzyme for its carbohydrate substrate has any practical importance, but it further adds to the picture of a tissue with a poorer ability to utilize circulating glucose than mammalian brain.

Part of the explanation for the failure of the locust nervous tissue to make effective use of trehalose except when it is present at concentrations greater than 50 mM, may lie in the kinetic characteristics of the enzyme trehalase. Although the maximum specific activity of the enzyme is apparently sufficient to supply adequate glucose to support the maximum

Table 6. K_m Values for Glucose of Hexokinase from Various Nervous Systems

Animal	Tissue	K_m (M)	References[a]
Locust	Thoracic ganglia	2.2×10^{-4}	1
Mouse	Brain	4.0×10^{-5}	2
Guinea pig	Brain	7.4×10^{-5}	3
Sheep	Brain	2.0×10^{-5}	4

[a] References: (1) Strang *et al.*, 1979; (2) Lowry *et al.*, 1964; (3) Bachelard, 1967; (4) Raggi and Kronfeld, 1966.

rate of oxygen consumption, the K_m of the enzyme for trehalose is 10 mM. As the concentration of the sugar in the vicinity of the enzyme falls below 50 mM it will perform less and less effectively. In addition, the concentration of trehalose in the ganglia both *in vitro* and *in vivo* is lower than that in the surrounding medium by a factor of from 5 to 8 (Strang and Clement, 1980). The same gradient applies in the case of the cockroach nervous system (Treherne, 1960). These two features, high K_m of the enzyme for its substrate and low tissue concentration of substrate, would seem adequate to account for the low utilization of trehalose, except for the fact that when intact ganglia are suspended *in vitro* in the presence of trehalose, there is a considerable efflux of glucose. In fact, the enzyme in intact tissue seems to be maximally active in the presence of 50 mM trehalose. As has been discussed elsewhere, a possible reason for this is that both enzyme and substrate occupy a relatively small compartment in the intact tissue (Strang and Clement, 1980).

The high concentration of trehalose in the hemolymph of the resting insect represents an immediately available reservoir of energy for the use of the wing muscles. Not only does the trehalase in the locust flight muscle have a maximum specific activity of 10 μmol/g/min at 37°C (Candy, 1974), but it has a K_m of 3 mM. Both these parameters indicate that the flight muscle is a more efficient user of trehalose than is the nervous system, especially as the concentration falls. But whereas the activity of the flight muscle trehalase is tightly controlled in the resting insect, that of the nervous system may be always fully active, implying a considerable difference in their kinetics.

5. Exogenous Substrates Other than Carbohydrate

If concentrations of carbohydrate fall too low, at times, to support aerobic metabolism in the nervous system of locusts and other insects, the corollary must be that there are at such times adequate alternatives, either exogenous or endogenous. There is definite proof that carbohydrate can support oxidative metabolism in the intact ganglia, but in the case of other substrates the evidence is mostly circumstantial.

5.1. Amino Acids

Although even within a single species the quantitative estimates of the amino acids in the hemolymph are as various as their estimators, two amino acids stand out as quantitatively important in the large orthopteran and dictyopteran species (as in many others, Jeuniaux, 1971). These two are glycine and proline. Although there are reports of high concentrations of glutamate, this amino acid is often estimated along with glutamine, leading to some confusion.

Glycine has not generally been considered as a potential energy substrate, but rather in the context of hemolymph osmoregulation, for which the insect has a considerable capacity (Berridge, 1970). Proline is a much more likely substrate for oxidation, especially by analogy with the flight muscles of various insects, such as the tsetse fly (Bursell, 1963), the blowfly (Sacktor and Childress, 1967), and the Colorado beetle (de Kort *et al.* 1973). (See also the chapter by E. Bursell in this volume.) In the orthopteran and dictyopteran species mainly under consideration here, the amino acid does not seem to play a very great part in the flight muscle metabolism (Mayer and Candy, 1969; Worm and Beenakkers, 1980). Thus, its concentration in the hemolymph of flying locusts falls either very slowly or not at all. This would certainly mean that it would be available to the nervous system at a time when the carbohydrate concentration is very low. The enzyme responsible for the initial metabolic step in the utilization of proline is proline dehydrogenase, which, in the thoracic ganglia of the locust *Schistocerca americana gregaria*, is present at a higher specific activity than in the flight muscle of the same species [respectively, 5.2 μmol/g/min at 37°C (Strang, R.H.C., unpublished observation) and 1.7 μmol/g/min at 25°C (Crabtree and Newsholme, 1970)].

In those tissues in which proline has been found to play a significant role in energy production, the main means of its utilization has been that found by Bursell (1965; this volume, chapter 5) in the flight muscle of the tsetse fly, which utilizes the citric acid cycle but does not fully oxidize the proline, which finally accumulates as alanine. In this pathway, a molecule of alanine is formed for each one of proline used. The pathway depends on the activity of L-alanine: oxoglutarate aminotransferase. An alternative mechanism involves the oxidative deamination of glutamate derived from proline, in which case the accumulation of ammonia is greater than that of alanine (Hansford and Johnson, 1975). In view of the wide differences in the estimated values of maximal specific activity of

Table 7. Maximal Specific Activities of Some Enzymes Associated with Amino Acid Metabolism in Various Nervous Tissues[a]

Enzyme	Tissue	Specific activity (μmol/g/min)	References[b]
Glutamate dehydrogenase	Locust thoracic ganglia	1.8	1
	Locust cerebral ganglia	23.3	2
	Rat cerebral cortex	0.82	1
	Mouse brain	24.3	2
	Rat brain	13.6	
Alanine: oxoglutarate aminotransferase	Locust cerebral ganglia	7.5	2
	Locust thoracic ganglia	7.4	1
	Cockroach cerebral ganglia	26.7	2
	Mouse brain	0.8	2
	Rat cerebral cortex	0.65	1
	Rat brain	1.5	2
Asparate: oxoglutarate aminotransferase	Locust thoracic ganglia	11.5	1
	Locust cerebral ganglia	50.8	2
	Rat brain	17.8	1
	Mouse brain	74.8	2
	Rat brain	48.0	2
Glutamine synthetase	Locust thoracic ganglia	1.1	1
	Rat cerebral cortex	2.4	1

[a] All the assays were performed at 25°C, and were made with crude homogenates.
[b] References: (1) Bradford *et al.*, 1969; (2) Sugden and Newsholme, 1975.

glutamate dehydrogenase by different authors, it is difficult to say what is the full potential of the enzyme in the insect nervous system; but it seems to be at least equal to that in the mammalian brain (Table 7). Both these mechanisms for the derivation of energy from amino acids (particularly proline) depend on the presence of an enzyme which catalyzes the conversion of oxaloacetate or malate to pyruvate, which may then be either transaminated or oxidized via the cycle. Possible enzymes for this purpose are oxaloacetate decarboxylase, found in the flight muscle of the tsetse fly (Bursell, 1965), and malic enzyme (Pearson *et al.*, 1979). Details of these enzymes are given in Table 8. Both enzymes are present in the thorax of the locust, but their exact location is not certain and is assumed to be flight muscle. Those muscles which are specialized for the use of proline as their main energy substrate are characterized by having high specific activities of malic enzyme, alanine:oxoglutarate aminotransferase, and proline dehydrogenase, and low activities of glycolytic enzymes, (Pearson *et al.*, 1979). It would be interesting to make the appropriate comparisons for the locust and cockroach nervous systems. The pathway may be complicated by compartmentation, as mitochondria are

Table 8. Possible Enzymes Involved in Amino Acid Metabolism in the Insect Nervous System

Enzyme (trivial and systematic name)	E.C. number	Reaction	$\Delta G^{0'}$ (kcal/mol)
Oxaloacetate decarboxylase [oxaloacetate carboxylase]	4.1.1.3	oxaloacetate ⇌ pyruvate + CO_2	?
Malic enzyme or malate dehydrogenase (decarboxylating) [L-malate:NAD(P) oxidoreductase (decarboxylating)]	1.1.1.38	malate + NAD(P) ⇌ pyruvate + NAD(P)H + CO_2	−0.36
Pyruvate carboxylase [pyruvate:carbon dioxide ligase (ADP)]	6.4.1.1	oxaloacetate + ADP + Pi ⇌ ATP + pyruvate + CO_2	+0.50

not generally permeable to oxaloacetate, and the enzyme oxaloacetate decarboxylase in the tsetse fly muscle has been reported to be located in the soluble fraction of the disrupted tissue (Bursell, 1965). If this is so, then some mechanism based on a shuttle would seem to be required.

Another possible enzyme which would allow the operation of such a pathway is pyruvate carboxylase (Table 8), which has been found in locust flight muscle (Crabtree *et al.*, 1972). As the known $\Delta G^{o'}$ figures for the reactions show, they are reversible and would proceed in the direction dictated by the relative concentrations of the reactants and products. Despite this, pyruvate carboxylase is generally considered to be an anaplerotic enzyme functioning to prime the citric acid cycle by the formation of oxaloacetate from pyruvate.

Although no direct test has been made of the ability of proline as an energy substrate in insect nerve cord, glutamate has been examined in this respect. Although this amino acid is present in the hemolymph of locusts in concentrations well below 1 mM (Mayer and Candy, 1969; Schlesinger *et al.*, 1977; Jabbar and Strang, unpublished observation) it is on the direct pathway of proline utilization, and so its oxidation is of obvious relevance for the use of the more abundant proline. At a concentration of 10 mM, it has been found to support a rate of oxygen uptake in isolated locust nerve cord slightly in excess of that found with 10 mM glucose. In addition, the finding that label from [^{14}C]glutamate accumulated in alanine would indicate the presence of the already discussed pathway (Bradford *et al.*, 1969). The low concentration of glutamate in the hemolymph, and the fact that, in the cockroach at least, there is a specific energy-requiring uptake mechanism that leads to a high concentration of glutamate within the nerve cord (Evans, 1975), argue against glutamate as an energy substrate in its own right. Another possibility is that glutamate and proline could have an anaplerotic function in energy metabolism, as is suggested by Worm and Beenakkers (1980) for locust flight muscle. Bradford *et al.* (1969), however, found no additive effect on oxygen uptake of glucose and glutamate when these were present together in the medium. In the experiments in question the tissue was unstimulated, and presumably working below its full metabolic capacity, so it is not certain that any anaplerotic effect of the amino acid would be apparent.

Attempts to follow stoichiometric relationships between amino acids in the isolated locust nervous system have been complicated by the large and rapid accumulation of alanine. Even under the best possible conditions *in vitro*, this far outweighs the decline of any other single free amino acid (Jabbar and Strang, unpublished observation). This rapid accumulation of alanine in insect tissues has been remarked on before (Price, 1961; Ray, 1964; Bradford *et al.*, 1969), and is particularly associated with anaerobic conditions. The most likely reason for it is the rapid accumulation of pyruvate resulting from increased glycolysis, which has been found to occur *in vitro* (Clement and Strang, 1978). In the locust nerve cord, alanine accumulation is a much more sensitive indicator of anaerobic conditions than is the accumulation of lactate. On a quantitative basis, about four to five times as much alanine as lactate accumulates during a 30 min incubation (Clement and Strang, 1978; A. Jabbar and R.H.C. Strang, unpublished observation). This ratio is almost exactly the same as that found for the proportional incorporation of label from [^{14}C] glucose in the same tissue under similar conditions (Bradford *et al.*, 1969). An interesting aspect of this is the fact that the maximal specific activity of lactate dehydrogenase of the thoracic ganglia (10 μmol/g/min at 37°C) is only slightly less than that of the alanine:oxoglutarate aminotransferase (7–8 μmol/g/min at 25°C). A possible explanation for this much more rapid accumulation of alanine than lactate could be that the enzymes are located within different cell types, with those containing the transferase having a higher glycolytic flux than those containing lactate dehydrogenase. Alternatively, the difference might be due to kinetic differences between the two enzymes with regard to substrate and inhibitor concentrations, as occurs in the marine worm *Arenicola marina*, in which the transferase has a K_m value for pyruvate 10-fold lower than the vertebrate enzyme (Felbeck and Grieshaber, 1980). Experiments with the locust thoracic ganglia *in vitro* over periods of up to one hour have demonstrated that while there is no gross increase in free amino nitrogen, there is a large transfer of such groups from many amino acids to pyruvate, with the formation of alanine, which has been found to diffuse easily out of the tissue. It is an interesting speculation that this function of alanine as a "sink" of amino nitrogen may occur *in vivo* in the nervous system under stressful conditions. Certainly glutamine formation is very low under these conditions, and if anything, the net flow of amino groups is from glutamine to alanine. There is, however, evidence that argues against this occurring *in vivo* from the work of Treherne (1960), who found little incorporation of label from radioactive glucose into alanine in the nervous system of living cockroaches, but a large accumulation in the amino acid when the experiment was done *in vitro*.

The formation of alanine under anaerobic conditions is not restricted to insects, but is a phenomenon found in many invertebrates (Stokes and Awapara, 1968). A fermentation mechanism to maintain the cytoplasmic redox potential was suggested to be present in insect flight muscle (Sacktor, 1965). It may be summarized thus:

$$\text{pyruvate} + \text{asparate} + \text{NADH} \rightleftharpoons \text{alanine} + \text{malate} + \text{NAD}$$

It is possible that the accumulation of alanine in the locust nervous system under anaerobic conditions is a result of this sort of mechanism. The finding of accumulations of malate or succinate would help to confirm the idea.

It is clear that there is the possibility in insect tissues of two mechanisms of amino acid metabolism, one aerobic and the other anaerobic, both leading to an accumulation of alanine. This complicates the interpretation of the gross amino acid metabolism of the ganglia *in vitro*, when both aerobic and anaerobic conditions could be present at the same time.

There is no such dramatic accumulation of alanine when mammalian nervous tissues are incubated. The probable reason for the difference between the two sorts of tissue is simply the different relative activities of the enzymes involved, lactate dehydrogenase and alanine:oxoglutarate aminotransferase (Tables 3 and 7).

5.2. Ketone Bodies and Lipids

By analogy with the mammal, ketone bodies are likely candidates as energy substrates for the nervous system. In mammals, they are utilized by the brain in relation to their concentration in the blood stream (Hawkins *et al.*, 1971), and it is they that provide the main source of energy for the brain during prolonged starvation (Owen *et al.*, 1967). In Table 9 are given the maximum specific activities of the enzymes associated with ketone body metabolism. Although the insects have lower activities than the mammals, the differences are not too great, and levels are sufficient to

Table 9. Maximal Activities of Enzymes Associated with Ketone Body Utilization in Various Tissues

Enzyme	Tissue	Temperature (°C)	Specific activity (μmol/g/min)	References[a]
β-hydroxy-butyrate dehydrogenase	Rat brain	37	1.18	1
	Rat brain	25	0.57	2
	Locust flight muscle	25	0.25	3
	Locust fat body	25	0.70	3
Acetoacetate CoA transferase	Locust cerebral ganglia	25	2.7	3
	Cockroach cerebral ganglia	25	4.0	4
	Mouse brain	25	9.8	4
	Rat brain	25	4.8	4
	Rat brain	25	2.1	2
Acetoacetyl CoA thiolase	Locust cerebral ganglia	25	1.2	4
	Cockroach cerebral ganglia	25	3.8	4
	Rat brain	25	2.1	2
	Rat brain	25	1.8	4
	Mouse brain	25	2.7	4

[a] References: (1) Cremer, 1971; (2) Williamson *et al.*, 1971; (3) Hill *et al.*, 1972; (4) Sugden and Newsholme, 1973.

support the maximum rates of oxygen uptake found *in vitro*. Differences appear however, when the concentration in the blood, and the increase due to starvation, are considered (Table 10). The total concentration in the hemolymph of resting locusts is about half that found in the blood of a fed rat, but while a 96 hour fast will raise the concentration in the rat blood by a factor of 15, there is only a doubling in the locust as a result of a 28 day fast. In humans, after starvation for 40 days, when ketone bodies are the main substrate for energy generation, the concentration of ketone bodies is 30 times the maximum found in the locust hemolymph. Although flying increases the concentration of ketone bodies, as might be expected from the use of lipid by the flight muscles, the increase is a modest one. Finally, rates of oxygen uptake *in vitro* for insect tissue using ketone bodies are very low, (Table 11), and in some tissues at least, actually decline during starvation. On the basis of these data, it seems unlikely that ketone bodies will make a great contribution to the metabolism of the nervous system of the locust. The question has still to be put to direct test.

In quantitative terms, the most obvious alternative substrate to car-

Table 10. Concentrations of Ketone Bodies in the Blood of Mammals and Insects under Different Conditions

		Concentration in blood (mM)			
Animal	Conditions	β-OH-butyrate	Acetoacetate	Total	References[a]
Locust	Fed	0.02	0.08	0.10	1
	3 day fast	0.09	0.07	0.16	1
	28 day fast	—	0.25	0.25	1
	1 hr flying	0.01	0.18	0.19	1
Man	40 day fast	6.67	1.17	7.48	2
Rat	Fed	0.09	0.13	0.22	3
	2 day fast	2.00	0.81	2.81	3
	4 day fast	2.62	0.73	3.35	3

[a] References: (1) Hill *et al.*, 1972; (2) Owen *et al.*, 1967; (3) Hawkins *et al.*, 1971.

bohydrate during flight and starvation is diglyceride, the concentration of which rises by threefold in the course of flying (Tietz, 1967; Mayer and Candy, 1969), when the concentration of carbohydrate is falling. No evidence is available about the capacity of the insect nervous system to hydrolyze the diglyceride to fatty acids and glycerol, and then to oxidize these materials at a sufficient rate. Glycerol itself rises to quite high concentrations in the blood of the flying locust, reaching a concentration of 4 mM after an hour of flight (Candy *et al.*, 1976), and so must be a potential substrate.

Table 11. Rates of Utilization of Ketone Bodies by Different Mammalian and Insect Tissues

Tissue	Conditions		Rate of utilization (μmol/g/min)	References[a]
Locust fat body	*In vitro* 37°C	Fed	0.007	1
		4-day fast	0.001	
Locust flight muscle	*In vitro* 37°C	Fed	0.040	1
		4-day fast	0.001	
Human brain	*In vivo*	40-day fast	0.240	2
Rat brain	*In vivo*	Fed	0.015	3
		2-day fast	0.100	

[a] References: (1) Hill *et al.*, 1972; (2) Owen *et al.*, 1967; (3) Hawkins *et al.*, 1971.

If it is found that the central nervous system of the locust and other insects can use fatty acids for their energy purposes, then they would be different from the mammalian brain, which cannot adequately use free fatty acids, at least *in vivo*.

6. Endogenous Substrates

6.1. Glycogen

6.1.1. General Metabolism

Among potential endogenous substrates, glycogen is paramount. In the cockroach nervous system, there is little evidence of significant deposits of fat (Wigglesworth, 1960). In the few insect nervous systems where it has been estimated, glycogen occurs in concentrations of from 11 to 60 μmol glucose/g in the ganglionic tissue (Treherne, 1960; Hart and Steele, 1973; Clement and Strang, 1978). This contrasts with the situation in the mammalian brain, in which the concentration is much lower (1–2 μmol glucose/g; Strang and Bachelard, 1973). The deposits in the insect nervous system are much more akin in amount to those found, for instance, in mammalian muscle, and not much lower than in the liver. In these tissues, it has a clearly defined role as an easily available store of energy. Interestingly, the sorts of concentration found in the locust and cockroach nerve cord are even higher than those in the flight muscle. Figures for the locust have been recorded to be as low as 7–8 μmol/g (Rowan and Newsholme, 1979) and 27–30 μmol/g (Worm and Beenakkers, 1980). (As the first of these figures refers only to the acid-soluble glycogen, it may be an underestimate.) The histochemical studies of Wigglesworth (1960) with the cockroach thoracic ganglia, indicated that there were heavy deposits of glycogen in the perineurial cells, and at the axonal roots of the neuronal cells. Wigglesworth showed that the deposits are sensitive to the concentration of the circulating carbohydrate, being depleted by starvation and augmented by the injection of glucose into the insect. In contrast, the glycogen of the mammalian brain is rather insensitive to such influences. This may simply be a reflection of the more exact glucose homeostasis in mammals, because insulin-induced hypoglycemia produces a fall in brain glycogen (Bachelard, 1976; Agardh *et al.*, 1978), as does ischemia (Kobayashi *et al.*, 1977). Thus, the brain glycogen concentration cannot be said to be completely immune to the concentration of circulating carbohydrate, but it certainly fluctuates within narrower limits than in the insect nerve cord.

Despite the obvious potential store of glycogen in the insect nervous system, hard evidence is lacking as to its exact function and the conditions in the living insect under which it is mobilized. Evidence from *in vitro* work suggests that in the absence of exogenous substrate, glycogen is the main source of energy for the tissue (Strang *et al.*, 1979). The cessation of spontaneous activity and the disappearance of high energy phosphates coincide with the reduction of glycogen to its minimim value. Assessing the role of glycogen *in vitro* is difficult because, even in the presence of exogenous glucose, there is a fall in glycogen concentration, although it is not so steep or so complete as in the absence of glucose (Clement and Strang, 1978). Another difficulty is that the rate of glycogenolysis exceeds that of glucose-6-phosphate oxidation. This leads to an efflux of carbohydrate in the form of glucose and trehalose from the tissue (Hart and Steele, 1973; Strang *et al.*, 1979). In the case of the first-named authors, a 1 hr incubation of the cockroach nerve cord in medium equilibrated only with the atmosphere produced an 84% fall in the concentration of glycogen from its initial level of 50–60 μmol glucose/g. Almost all of this was lost to the tissue mainly in the form of trehalose. When locust tissue was incubated in a medium equilibrated with 100% oxygen, the loss was only 30% of the hexose formed. The free glucose released is unlikely to have come from glucose-6-phosphate, as the tissue is devoid of glucose-6-phosphatase activity (Strang *et al.*, 1979). Much more likely is that it came from the action of trehalase, although glycogen debranching enzyme might have produced some also.

The loss of glucose and trehalose from the nervous system *in vitro* seems unlikely to have a counterpart in the living insect, and it adds to the difficulty of assessing the role of glycogen in the ganglia *in vivo*. To determine this, it will be required to find under what conditions of stress (e.g., flying) the glycogen of the central nervous system is depleted, and what is happening simultaneously to the carbohydrate in other tissues. Does it, for instance, fall in sequence after the glycogen of the flight muscle and the fat body?

Another difference between the glycogen of the nervous systems of the mammal and the locust of which the meaning and the mechanism are unclear, is the relationship between the concentrations of glycogen and free glucose in the stressed and isolated tissues. In the decapitated mouse, for instance, glucose is rapidly depleted, and only when it has gone does the concentration of glycogen decline (Lowry *et al.*, 1964). This observation, along with that of the low concentration of brain glycogen, have led to the idea that it is a store of energy for use only in extremity. The situation in the locust thoracic ganglia *in vitro* seems to be different. Under both aerobic and anaerobic conditions, in the absence of exogenous substrate, there is a rapid decline in the concentration of glycogen, but not in that of glucose (Strang *et al.*, 1979). At the moment, it is impossible to say whether this effect is due to inhibition of glucose utilization, or to its replacement by the hydrolysis of trehalose or glycogen, or a combination of both these mechanisms.

It is also unclear if this is likely to take place *in vivo*, and if it does, then what might be its significance. Central to the question are the kinetics of hexokinase. In common with the hexokinase of other tissues, including that of the mammalian cerebral cortex, that of the locust thoracic ganglia is inhibited by glucose-6-phosphate (Flanigan, T. and Strang, R.H.C., unpublished observations). Although the degree of inhibition at comparable concentrations is less in the case of the locust, [0.25 mM glucose-6-phosphate produced an inhibition of 35% with concentrations of glucose ranging from 0.4 to 2.0 mM, compared to an inhibition of 80% produced by the same concentration in the case of the soluble enzyme from the guinea pig cerebral cortex (Newsholme *et al.*, 1968)], in view of the lower maximum specific activity of the locust enzyme, the effect of the inhibition might be to limit the utilization of free glucose. Indeed, it may be possible to explain the consistency of glucose concentration under stressful conditions on this basis. In the mammal, factors such as ischemia, which produce increased glycolysis, result in a decline in the concentration of glucose-6-phosphate (Lowry *et al.*, 1964; Rolleston and Newsholme, 1967). This is not true in other species such as turtle, fish, and frog, in which decapitation causes a rise in the concentration of glucose-6-phosphate to above 1 mM, which in turn results in a decline in the rate of glucose utilization in the tissue (McDougal *et al.*, 1968). Interestingly, at the time when the locust starts to fly, the concentration of glucose-6-phosphate in the flight muscle increases sharply from concentrations of 0.1–0.15 μmol/g to 0.25–0.3 μmol/g (Rowan and Newsholme, 1979; Worm and Beenakkers, 1980). All these tissues discussed have in common large deposits of glycogen (10–60 μmol glucose/g), which are rapidly mobilized. Maybe a similar effect is present in the isolated thoracic ganglia. Further speculation is pointless, in the absence of precise information about the concentration of glucose-6-phosphate in the tissues under different conditions *in vivo* and *in vitro*.

6.1.2. Phosphorylase

Returning to the question of the control of glycogen metabolism, the important enzyme in this regard is likely to be phosphorylase. A large experimental effort over many years and involving many tissues, has indicated that this enzyme is a most complex point of control, subject to a variety of metabolic and hormonal factors. The paradigm, derived, for instance, from mammalian skeletal muscle, is that of two forms of the enzyme, a and b, with the latter form inactive except in the presence of high concentrations of AMP. Conversion of the b to the a form is effected by a kinase, whose activity is controlled by the concentration of cAMP. Investigation of the kinetics of the enzyme in mammalian brain revealed that the active form has a lower affinity for glycogen than does the enzyme from skeletal muscle; a fact that was thought to be able to explain to some extent the stability of the brain's glycogen stores (Lowry *et al.*, 1967). Childress and Sacktor (1970), in their meticulous examination of the kinetics of the phosphorylase from the blowfly flight muscle, separated the two forms of the enzyme and found a picture similar in qualitative terms to that familiar from mammalian tissues. The various substrates affected each other's binding to the enzyme. From a careful study of the concentration effects, they were able to estimate the proportions of the a and b forms in the tissues in resting and flying insects. Nothing so exact has been attempted with the insect nervous system, and in view of the Herculean task of collecting sufficient tissue, (Childress and Sacktor used 300 g of blowfly thorax in a typical experiment), perhaps it never will be. So far, the work done has been limited to considerations of the total potential activity and the proportions of the different forms of the enzyme in crude homogenates of nervous tissue. The two estimates of the maximal specific activity of phosphorylase for flight muscle of *Schistocerca* are in good agreement: 12.3 μmol/g/min at 37°C (Strang, unpublished observations), and 7.5 μmol/g/min at 25°C (Crabtree and Newsholme, 1972). As is the case with other enzymes concerned with energy metabolism, the full activity of phosphorylase in the nerve cord is lower than in the flight muscle; 3–6 μmol/g/min at 37°C (Strang, unpublished observations). Of possible ligands which might affect the kinetics of the phosphorylase, only AMP and glucose-6-phosphate have been examined. Unpublished work of the author has found that the sugar phosphate has no apparent effect upon the full activity of the phosphorlase in homogenates of the locust thoracic ganglia, unlike the situation in flight muscle assayed simultaneously, in which the phosphate inhibits both the full activity and that found in the absence of AMP. If glucose-6-phosphate has no effect on the turnover of glycogen in the locust nervous system, it would be consistent with the rapid breakdown found under various conditions; but in view of the complexity of the enzyme's kinetics, it is too early to make dogmatic statements. For one thing, some attempt to remove low-molecular-weight effectors is almost an obligatory requirement in work with phosphorylase, and this has not been attempted with the nerve cord enzyme. Moreover, when the enzyme is assayed, as it often is, by the release of inorganic phosphate from glucose-1-phosphate, the initial concentrations of the phosphate, much lower than those in the tissue, may distort the kinetics.

If glucose-6-phosphate exerts no control over the activity of the phosphorylase, then either other factors such as glucose or ATP must inhibit it, or the turnover of glycogen in the nervous system is very rapid. If the latter were true, the specific activity of glycogen synthetase in the intact tissue must be equal to or greater than the activity of the phosphorylase, under normal conditions. Nothing is known of the absolute rates of glycogen synthetase in the nerve cords of orthopterans. Experiments on the incorporation of label from [^{14}C]glucose do not support the theory of an especially rapid turnover. Isolated locust thoracic ganglia incubated under the best conditions of substrate and oxygen, incorporated glucose at a rate of 4–6 nmol/g/min at 37°C (Powell and Strang, unpublished observations). In almost exactly comparable experiments with slices of guinea pig cortex, the rate was 0.8 nmol/g/min (Quach *et al.*, 1978). Considering the difference in total concentration of glycogen between insect and mammalian nervous tissue, the former does not display a particularly high rate of incorporation.

6.1.3. Humoral Factors Affecting Phosphorylase Activity

The possession of a system of interconvertible active and inactive enzymes is an evolutionary sophistication, imposed on the control exerted by metabolic ligands. It allows an anticipatory change, or a change due to physiological rather than biochemical events. It is reasonable to expect to find the mechanism in tissues with large stores of glycogen and which may require a lot of energy quickly. The phosphorylase of the flight muscle and fat body of the insect show interconvertable forms of phosphorylase, although in the former, the conversion is likely to be mediated by Ca^{2+} and inorganic phosphate (Sacktor *et al.*, 1974) and in the latter by a hormone (Steele, 1963). In view of the high concentrations of glycogen in the insect nervous system, it is reasonable to assume that the mechanism might also exist in that tissue. In the main, two humoral factors have been examined with a view to their physiological function in activating phosphorylase by converting phosphorylase b to phosphorylase a; hyperglycemic hormone and octopamine. The first of these was found to have the effect of increasing the concentration of circulating trehalose in the cockroach as a result of glycolgenolysis in the fat body. It increased the ratio of active to inactive forms of phosphorylase in that tissue (Steele, 1963). Despite the clear evidence of its action, the physiological role of the hyperglycemic hormone is still quite uncertain, especially in view of the fact that it requires a few hours to achieve its maximal effect on hemolymph carbohydrate (Steele, 1976). Steele, (1963) found that although the hormone had no effect on the flight muscle tissue, it did deplete the glycogen of the nerve cord. This work was later extended, and Hart and Steele (1973) found that injection of the extract of corpus cardiacum into the insect had the effect of reducing the concentration of nerve cord glycogen by an average of 42 μmol glucose/g in 30 min. Simply averaged over the period, this would produce a rate of production of 1.4 μmol hexose-phosphate/g/min, more than double the maximum rate of glucose consumption found *in vitro* on the basis of O_2 consumption. Under these conditions, it seems likely that much of the released carbohydrate might be lost from the tissue, unless the concentration of glucose and trehalose in the hemolymph were also high. One of the difficulties in reproducing this glycogenolytic effect *in vitro* is the rapid decline and almost total disappearance of glycogen in artificial media.

It is probable, but not certain, that the hormone has its effect via an increase in the concentration of cAMP, which in turn converts phosphorylase b into phosphorylase a. As with so many other aspects of insect metabolism, the crucial measurements have not been made: in this case, the activities of adenyl cyclase and cAMP in the tissues in the presence of the hormone. Until this has been done, the final link between cause and effect will be lacking.

A puzzling aspect of the effect of the extracts of the corpus cardiacum on the glycogenolysis of the nerve cord is that they are nullified by serotonin (5-hydroxy tryptamine) at a concentration of 10^{-3} M (Hart and Steele, 1973). Serotonin by itself causes a deactivation of phosphorylase activity by increasing the proportion of phosphorylase b (Hart and Steele, 1969). This is puzzling, because serotonin also has been found to activate adenyl cyclase in the cockroach and locust ganglia (Nathanson and Greengard, 1973; Kilpatrick *et al.*, 1980). The paradox is more apparent than real, however, as the amine has its maximal effect on the adenyl cyclase at a concentration of 10^{-5} M. At 10^{-3} M no such effect is apparent, and this removes the contradiction in the results. This does not explain the antagonistic effect to hyperglycemic hormone, which must operate through some other mechanism, such as those discussed by Steele (1976). Again, the key measurement, the cellular concentration of cyclic AMP, has not been made. It would be of great interest not only to measure the cyclic nucleotide in the tissue as a whole, but to use a histochemical method to determine its cellular distribution.

Another currently favorite candidate as a humoral effector of glycogenolysis in the nerve cord is octopamine. The enthusiasm is such that this amine, which is certainly present in the insect nervous system (Robertson and Steele, 1974; Evans, 1978a), is in some danger of being cast in the role of general factotum. It has been found to cause a pronounced enhancement of the activity of adenyl cyclase in the nerve cord of cockroach and locust on both intact tissue and homogenates. In both of these insects, it produced its greatest effect at a concentration of 250 μM (Nathanson and Greengard, 1973; Kilpatrick *et al.*, 1980). Robertson and Steele (1972) found a 33% decrease in nerve cord glycogen in two hours under the influence of a similar concentration of octopamine, and this was accompanied by an activation of phosphorylase by an increase in the proportion of the active form of the enzyme. Effects on glycogen metabolism were found with concentrations as low as 500 nM. Downer, (1979) reported a hypertrehalosemic effect of octompamine in the cockroach at much lower concentrations of octopamine (10 nM) and found that the effect was quite independent of effects produced by extracts of corpus cardiacum. Candy (1978) has found that the amine at a concentration of 1 μM in flight muscle perfusates stimulated both carbohydrate metabolism and the strength of the muscular contractions. Other reported physiological effects have been on nervous conduction (Hoyle, 1975) and heart beat of the cockroach (Collins and Miller, 1977). In all these reported experiments in which octopamine is either injected into the hemocoel or perfused over preparations *in vitro*, the implicit assumption is that the amine acts as a humoral agent in the living insect, by analogy with the situation in the lobster, in which it is definitely located in, and released from, the neurosecretory organ (Evans *et al.*, 1975). However, the distribution of the amine may be quite different in the two invertebrates: peripheral in the lobster, and central in the insect (Wallace *et al.*, 1974). The keystone that would unite and give stability to the demi-arches of physiology and biochemistry is the concentration of circulating octopamine under different conditions. In his contribution to this volume, Candy reports a steep increase in the concentration of circulating D-octopamine in the locust hemolymph when the insect starts to fly. The concentration reaches a maximum of between 150–200 nM. Under these conditions, the octopamine is acting in the manner expected of a humoral effector. Nevertheless, the concentration found is less than that of 500 nM required to produce an effect on nerve cord glycogen.

In discussing these two hyperglycemic factors of the insect, it is difficult to escape the analogy between hyperglycemic hormone and mammalian glucagon, both peptides and slow in action; and between octopamine and adrenaline, both derived from catecholamines and rapidly acting in their glycogenolytic effect. If the insect factors do act on the glycogen of the nerve cord under natural conditions, it would mark another difference between the insect and mammalian nervous tissues. The lack of effect of adrenaline on the brain glycogen of higher species may be more a physiological than biochemical phenomenon, as the neonatal chicken brain will respond to the effect of adrenaline before the blood–brain barrier is fully formed (Edwards *et al.*, 1974). This, in turn, would indicate that the blood–brain barrier of the insect nervous system is different in this respect from that of the mammal, or that the effect of the factors was on the superficial cells of the perineurium, which might lie outside such a barrier. Recently, Steele and Chan (1980) have found a strong metabolic response to octopamine in terms of increased oxygen uptake by the isolated cockroach thoracic ganglia, which they consider to be associated with a Na^+ uptake mechanism in the perineurial cells. It is these cells that contain large deposits of glycogen (Wigglesworth, 1960). Steele and Chan suggest that the glycogenolysis could result from purely metabolic changes. A complicating factor in interpreting this effect of octopamine is the finding by Evans (1978b) that there is a specific uptake mechanism for octopamine into the cockroach nervous system dependent upon the presence of Na^+ in the medium. It will be difficult to disentangle these two effects, both of which will have some effect on metabolic activity.

7. Conclusion

I have tried where possible to make comparisons between the metabolism of the thoracic and cerebral ganglia of the cockroach and locust and the flight muscles of those and other insects; and also with the mammalian brain. The metabolic activity of the nervous tissue in terms of maximal specific activities of rate-limiting enzymes and of recorded oxygen consumption is only a fraction of that of the locust flight muscles. This is consistent with insect flight muscle in action having the most vigorous aerobic metabolism of any known tissue. On the other hand, the metabolism of the flight muscles at rest is very tightly controlled, and in this may contrast with the nerve tissue. Trehalase may offer an example of this. The enzyme in the flight muscle is practically inactive until flight (Van der Horst *et al.*, 1978). If the results from the intact ganglia *in vitro* represent the conditions in the living animal, the trehalose of that tissue may always be fully active. As with so many features of insect metabolism, little is known about the control of this enzyme.

The question of how closely the conditions of the tissues *in vitro* reflect those *in vivo* is central to the interpretation of many of the results presented. It is inevitable that a great deal of biochemical work will be done with ganglia outside the body. This ensures that the tissue is free of other contaminating tissues, and that a collection of ganglia from different insects are all subjected to the same conditions; an important consideration where there is such a wide natural variation among different individuals of the same species. In many respects the state of the tissues under optimum conditions outside the insect does reflect that in the intact animal. But this is not uniformly true. Changes in the metabolism of glycogen and amino acids are sensitive indicators of deficient oxygen supply, and even concentrations of the gas exceeding 700 μM in artificial media, do not fully compensate for the loss of the tracheal system. Deductions from experiments *in vitro* must be carefully tempered by measurements in the living insect when possible.

The comparison of the insect central nervous system with insect flight muscle is an immediate and practical one. They occur in the same body. The comparison with the mammalian brain is a purely intellectual exercise, but may cast some light on the biology of the different classes. The outstanding characteristic of the energy metabolism of the mammalian brain is its normal dependence on, and specialization in the use of, an adequate concentration of circulating glucose. Probably this is possible only as a result of the careful homeostasis of the blood glucose by a variety of hormonal and metabolic mechanisms. The mammalian brain has little store of endogenous substrate. If we assume (and there is really no evidence for it) that the main energy substrate of the nervous system of the locust and cockroach at rest is carbohydrate, the evidence suggests its capacity to use this material is less than that of the mammalian brain. The hypothesis that has directed much of the author's work is that under some conditions such as flying, carbohydrate concentration in the hemolymph ceases to be an adequate substrate for the nervous system for some insects. Definite proof of this is still lacking. The capacity of the nervous system to utilize other exogenous substrates is mostly untested. Although the concentration of glycogen in the ganglia far exceeds that in the mammalian brain, the conditions of its mobilization and utilization are uncertain, and the results from work *in vitro* may be misleading.

In this sketch map of aspects of the metabolism of the insect nervous system there are many blanks. It is these blanks that the old mapmakers filled with drawings of dragons and anthropophagi. In their place I have substituted speculation. The monsters and speculation should be treated with equal seriousness. Nevertheless, both have the function of beckoning to the potential explorer and saying, "Come and prove me wrong!"

I hope that I have not slighted anyone else's work in this chapter by either omission or commission. If I have, it is due to ignorance, and not intention.

References

Agardh, C.D., Folbergrova, J., and Siesjo, B.K., 1978, Cerebral metabolic changes in profound insulin-induced hypoglycaemia and in the recovery period following glucose administration, *J. Neurochem.* **31:**1135.

Alp, P., 1973, M.S. thesis, Oxford University.

Bachelard, H.S., 1967, The subcellular distribution and properties of hexokinase in the guinea-pig cerebral cortex, *Biochem. J.* **104**:286.

Bachelard, H.S., 1976, Metabolic and deficiency diseases of the nervous system, in: *Handbook of Chemical Neurology*, Volume 27 (P.J. Vinken and G.W. Bruyn, eds.), Part 1, p. 7, North-Holland, Amsterdam.

Bachelard, H.S., 1978, Glucose as a fuel for the brain, *Biochem. Soc. Trans.* **6**:520.

Balázs, R. Kovaćs, S., Teichgraber, P. Cocks, W.A., and Eayrs, J.T., 1968, Effects of neonatal thyroid deficiency in the developing brain, *J. Neurochem.* **15**:1335.

Banks, P., Bartley, W., and Birt, L.M., 1968, in *The Biochemistry of the Tissues*, 2nd edition, p. 126, John Wiley and Sons, London.

Bedford, J.J., 1977, The carbohydrate levels of insect haemolymph, *Comp. Biochem. Physiol.* **57**:83.

Berridge, M.J., 1970, Osmoregulation in terrestrial arthropods, in: *Chemical Zoology*, Volume V (M. Florkin and B.J. Scheer, eds.), p. 290, Academic Press, New York and London.

Bradford, H.F., Chain, E.B., Cory, H.T., and Rose, S.P.R., 1969, Glucose and amino acid metabolism in some invertebrate nervous systems, *J. Neurochem.* **16**:969.

Bursell, E., 1963, Aspects of the metabolism of amino acids of the tsetse fly, *Glossina* (diptera), *J. Insect Physiol.* **9**:439.

Bursell, E., 1965, Oxaloacetate carboxylase in flight musculature of the tsetse fly, *Comp. Biochem, Physiol.* **16**:259.

Candy, D.J., 1974, The control of muscle trehalase activity during locust flight, *Biochem. Soc. Trans.* **2**:1107.

Candy, D.J., 1978, The regulation of locust flight muscle metabolism by octopamine and other compounds, *Insect Biochem.* **8**:177.

Candy D.J., Hall, L.J., and Spencer, I.M., 1976, The metabolism of glycerol in the locust *Schistocerca gregaria* during flight, *J. Insect Physiol.* **22**:583.

Childress, C.C., and Sacktor, B., 1970, Regulation of glycogen metabolism in insect flight muscle, *J. Biol. Chem.* **245**:2927.

Church, N.S., 1960, Heat loss and the body temperature of flying insects, *J. Exp. Biol.* **37**:171.

Clement, E.M., 1979, Some studies on the carbohydrate metabolism and electrophysiology of the locust (*Schistocerca gregaria*) nervous system *in vitro*, M.S., thesis, University of Glasgow.

Clement, E.M., and Strang, R.H.C., 1978, A comparison of some aspects of the physiology and metabolism of the nervous system of the locust *Schistocerca gregaria in vitro* with those *in vivo*, *J. Neurochem.* **31**:135.

Collins, C., and Miller, T., 1977, Studies in the action of biogenic amines on the cockroach heart, *J. Exp. Biol.* **67**:1.

Crabtree, B., and Newsholme, E.A., 1970, The activities of proline dehydrogenase, glutamate dehydrogenase, asparate-oxoglutarate aminotransferase and alanine-oxoglutarate aminotransferase in some insect flight muscles, *Biochem. J.* **117**:1019.

Crabtree, B. and Newsholme, E.A., 1972, The activities of phosphorylase, hexokinase, phosphofructokinase, lactate dehydrogenase and glycerol 3-phosphate dehydrogenase in muscles from vertebrates and invertebrates, *Biochem. J.* **126**:49.

Crabtree, B., Higgins, S.J., and Newsholme, E.A., 1972, The Activities of pyruvate decarboxylase, phosphoenolpyruvate carboxylase and fructose diphosphatase in muscles from vertebrates and invertebrates, *Biochem. J.* **130**:391.

Cremer, J.E. 1971, Incorporation of label from D-β-hydroxy (^{14}C) butyrate and (3-^{14}C) acetoacetate into amino acids in rat brain tissue, *in vivo*, *Biochem. J.* **122**:135.

Crone, C., 1965, Facilitated transfer of glucose from blood into brain tissue, *J. Physiol.* **181**:103.

Davis, R.A., and Fraenkel, G., 1940, The oxygen consumption of flies during flight, *J. Exp. Biol.* **17**:402.

de Kort, C.A.D., Bartelink, A.K.M., and Schuurmans, R.R., 1973, The significance of L-proline for oxidative metabolism in the flight muscles of the colorado beetle, *Leptinotarsa decemlineata*, *Insect Biochem.* **3**:11.

Downer, R., 1979, Induction of hypertrehalosemia by excitation in *Periplaneta americana*, *J. Insect Physiol.* **25**:59.

Edwards, C., Nahorski, S.R., and Rogers, K.J., 1974, *In vivo* changes of cerebral cyclic adenosine 3′, 5′ monophosphate induced by biogenic amines associated with phosphorylase activation, *J. Neurochem.* **22**:565.

Edwards, H.T., and Margaria, R., and Dill, D.E. 1934, Metabolic rate, blood sugar and the utilization of carbohydrate, *Am. J. Physiol.* **108**:203.

Elliot, K.A.C., 1955, Brain tissue respiration and glycolysis, in: *Neurochemistry* (K.A.C. Elliot, I. Page, and J.H. Quastel, eds.), p. 91, Thomas, Springfield.

Evans, P.D., 1975, The uptake of L-glutamate by the central nervous system of the cockroach *Periplaneta americana*, *J. Exp. Biol.* **62**:55.

Evans, P.D., 1978a, Octopamine distribution in the insect nervous system, *J. Neurochem.* **30**:1009.

Evans, P.D., 1978b, Octopamine: A high affinity uptake mechanism in the nervous system of the cockroach, *J. Neurochem.* **30**:1015.

Evans, P.D., Talamo, B.R., and Kravitz, E.A., 1975, Octopamine neurones: Morphology, release of octopamine and possible physiological role, *Brain Res.* **90**:340.

Felbeck, H., and Grieshaber, M.K., 1980, Investigations on some enzymes involved in the anaerobic metabolism of amino acids in *Arenicola marina*, L., *Comp. Biochem. Physiol.* **66B**:205.

Fishman, R.A., 1964, Carrier transport of glucose between blood and cerebrospinal fluid, *Am. J. Physiol.* **206**:836.

Florkin, M., and Jeuniaux, C., 1974, Haemolymph composition, in: *The Physiology of Insecta*, 2nd edition, Volume V (M. Rockstein, ed.), p. 283, Academic Press, New York.

Hansen, O., 1964, Effect of diet on the amount and composition of locust blood carbohydrate, *Biochem. J.* **92**:333.

Hansford, R.G., and Johnson, R.N., 1975, The nature and the control of the tricarboxylic acid cycle in beetle flight muscles, *Biochem, J.* **148**:389.

Härkönen, M.H.A., Passonneau, J.V., and Lowry, O.H., 1969, Relationships between energy reserves and function in rat superior cervical ganglion, *J. Neurochem.* **16**:1439.

Hart, D.E., and Steele, J.E., 1969, Inhibition of insect nerve cord phosphorylase activity by 5-hydroxytryptamine, *Experientia* **25**:243.

Hart, D.E., and Steele, J.E., 1973, The glycogenolytic effect of corpus cardiacum on the cockroach nerve cord, *J. Insect Physiol.* **19**:927.

Hawkins, R.A., Williamson, D.H., and Krebs, H.A., 1971, Ketone-body utilization by adult and suckling rat brain *in vivo*. *Biochem, J.* **122**:13.

Hermansen, L., Hultman, E., and Saltin, B., 1967, Muscle glycogen during prolonged severe exercise, *Acta Physiol. Scand.* **71**:129.

Hill, L., Izatt, M.E.G., Horne, J.A., and Bailey, E., 1972, Factors affecting concentrations of acetoacetate and D-3-hydroxybutyrate in the haemolymph and tissues of the adult desert locust, *J. Insect Physiol.* **18**:1265.

Hirsch, H.E., Blanco, C.E., and Parks, M.E., 1980, Glycerol phosphate dehydrogenase activity in multiple sclerosis plaques confirms localization in oligodendrocytes, *J. Neurochem.* **34**:760.

Horowicz, P., and Larrabee, M.G., 1957, Glucose consumption and lactate production in a mammalian sympathetic ganglion at rest and in activity, *J. Neurochem.* **2**:101.

Howden, G.F., and Kilby, B.A., 1956, Trehalose and trehalase in the locust, *Chem. Ind. (London)* **956**:1453.

Hoyle, G., 1975, Evidence that insect dorsal unpaired median neurones are octopaminergic, *J. Exp. Zool.* **193**:425.

Jeuniaux, C., 1971, Haemolymph—arthropoda, in: *Chemical Zoology*, Volume VI (M. Florkin and B.T. Scheer, eds.), p. 98, Academic Press, New York and London.

Johnson, M.K., 1960, The intracellular distribution of glycolytic and other enzymes in rat brain homogenates and mitochondrial preparations, *Biochem. J.* **77**:610.

Jutsum, A.R., and Goldsworthy, G.J., 1976, Fuels for flight in Locusta, *J. Insect Physiol.* **22**:243.

Kobayashi, M., Lust, W.D., and Passonneau, J.V., 1977, Concentrations of energy metabolites and cyclic nucleotides during and after bilateral ischaemia in the gerbil cerebral cortex, *J. Neurochem.* **29**:53.

Kilpatrick, A.T., Vaughan, P.F.T., and Donnellan, J.F., 1980, Monoamine sensitive adenylate cyclase in *Schistocerca gregaria* nervous tissue, in: *Insect Neurobiology and Pesticide Action* (M. Sherwood, ed.), p. 341, Society of Chemical Industry, London.

Larrabee, M.G. 1958, Oxygen consumption of excised sympathetic ganglia at rest and in activity, *J. Neurochem* **2**:81.

Lehninger, A.L., 1975, *Biochemistry*, 2nd edition, p. 456, Worth, New York.

Lowry, O.H., Passonneau, J.V., Hasselberger, F.X., and Schulz, D.W., 1964, Effect of ischaemia on known substrates and coafactors of the glycolytic pathway in brain, *J. Biol. Chem.* **239**:18.

Lowry, O.H., Schulz, D.W., and Passonneau, J.V., 1967, The kinetics of glycogen phosphorylase from brain and muscle, *J. Biol. Chem.* **242**:271.

McDougal, D.B., Holowach, J., Howe, M.C., Jones, E.M., and Thomas, C.A., 1968, The effects of anoxia upon energy sources and selected metabolic intermediates in the brains of fish, frog and turtle, *J. Neurochem.* **15**:577.

McIlwain, H., 1953, Glucose level, metabolism and response to electrical impulses in cerebral tissues from man and animals, *Biochem. J.* **55**:618.

McIlwain, H., and Bachelard, H.S., 1971a, *Biochemistry and the Central Nervous System*, 4th edition, p. 69, Churchill Livingstone, Edinburgh and London.

McIlwain, H., and Bachelard, H.S., 1971b, *Biochemistry and the Central Nervous System*, 4th edition, p. 160, Churchill Livingstone, Edinburgh and London.

McIlwain, H. Anguiano, G., and Cheshire, J.D., 1952, Electrical stimulation *in vitro* of the metabolism of glucose by mammalian cerebral cortex, *Biochem. J.* **50**:12.

Mayer, R.J., and Candy, D.J., 1969, Changes in energy reserves during flight of the desert locust *Schistocerca gregaria*, *Comp. Biochem. Physiol.* **31**:409.

Mwangi, R.W., and Goldsworthy, G.J., 1977, Interrelationships between haemolymph lipid and carbohydrate during starvation in *Locusta*, *J. Insect Physiol.* **23**:1275.

Nathanson, J.A., and Greengard, P., 1973, Octopamine-sensitive adenyl cyclase: Evidence for a biological role of octopamine in nervous tissue, *Science* **180**:308.

Newsholme, E.A., and Start, C., 1974, *Regulation in Metabolism*, pp. 133–135, John Wiley and Sons, London.

Newsholme, E.A., Rolleston, F.S., and Taylor, K. 1968, Factors affecting the glucose 6-phosphate inhibition of hexokinase from the cerebral cortex tissues of the guinea-pig, *Biochem. J.* **106**:193.

Newsholme, E.A., Sugden, P.H., and Williams, T., 1977, Effect of citrate on the activities of 6-phosphofructokinase from the nervous and muscle tissue from different animals, and the relationship to the regulation of glycolysis, *Biochem, J.* **166**:123.

Owen, O.E., Morgan, A.P., Kemp, H.G., Sullivan, J.M., Herrera, M.G., and Cahill, G.F., 1967, Brain metabolism during fasting, *J. Clin. Invest.* **46:**1589.
Pearson, D.J., Imbuga, M.O., and Hoek, J.B., 1979, Enzyme activities in flight and lef muscles of the dung beetle in relation to proline metabolism, *Insect Biochem.* **9:**461.
Poláček, I., and Kubišta, V., 1960, Metabolism of the cockroach *Periplaneta americana* during flight, *Physiol. Bohemoslov.* **9:**228.
Price, G.M., 1961, The accumulation of α-alanine in the housefly *Musca vicina*, *Biochem. J.* **81:**15P.
Quach, T.T., Rose, C., and Schwartz, J.C., 1978, [^{3}H]glycogen hydrolysis in brain slices: Response to neurotransmitters and modulation of noradrenaline receptors, *J. Neurochem.* **30:**1335.
Raggi, F., and Kronfeld, D.S., 1966, Higher glucose affinity of hexokinase in sheep brain than in rat brain, *Nature (London)* **209:**1353.
Ray, J.W., 1964, The free amino acid pool of the cockroach (*Periplaneta americana*) central nervous system and the effect of insecticides, *J. Insect Physiol.* **10:**587.
Robertson, H.A., and Steele, J.E., 1972, Activation of insect nerve cord phosphorylase by octopamine and adenine 3′,5′-monophosphate, *J. Neurochem.* **19:**1603.
Robertson, H.A., and Steele, J.E., 1974, Octopamine in the insect central nervous system: Distribution, biosynthesis and possible physiological role, *J. Physiol.* **237:**34P.
Robinson, N.L., and Goldsworthy, G.J., 1976, Adipokinetic hormone and flight metabolism in the locust, *J. Insect Physiol.* **22:**1559.
Robinson N.L., and Goldsworthy, G.J., 1977a, Adipokinetic hormone and the regulation of carbohydrate and the lipid metabolism in a working flight muscle preparation, *J. Insect Physiol.* **23:**9.
Robinson, N.L., and Goldsworthy, G.J., 1977b, A possible site of action of adipokinetic hormone on the flight muscle of locusts, *J. Insect Physiol.* **23:**152.
Rolleston, F.S., and Newsholme, E.A., 1967, Effects of fatty acids, ketone bodies, lactate and pyruvate on glucose utilization by guinea-pig cerebral cortex slices, *Biochem. J.* **104:**519.
Rowan A.N., and Newsholme, E.A., 1979, Changes in the contents of adenine nucleotides and intermediates of glycolysis and the citric acid cycle in flight muscles of the locust upon flight, and their relations to the control of the cycle, *Biochem, J.* **178:**209.
Sacktor, B., 1965, Energetics and respiratory metabolism of muscular contraction, in: *The Physiology of Insecta*, Volume II (M. Rockstein, ed.), p. 484, Academic Press, New York.
Sacktor, B., and Childress, C.C., 1967, Metabolism of proline in insect flight muscle and its significance in stimulating the oxidation of pyruvate, *Arch. Biochem. Biophys.* **120:**583.
Sacktor, B., Wu, N.-C., Lescure, O., and Reed, W.D., 1974, Regulation of muscle phosphorylase b kinase activity by inorganic phosphate and calcium ions, *Biochem. J.* **137:**535.
Schlesinger, H.M., Applebaum, S.A., and Birk, Y., 1977, Comparative uptake, tissue permeability and turnover of α,γ-diaminobutyric acid and 3-cyano-L-alanine in locusts, *J. Insect Physiol.* **23:**1311.
Sols, A., and Crane, R.K., 1954, The inhibition of brain hexokinase by adenosine diphosphate and sulphydral reagents, *J. Biol. Chem.* **206:**925.
Steele, J.E., 1963, The site of action of insect hyperglycaemic hormone, *Gen. Comp. Endocrinol.* **3:**46.
Steele, J.E., 1976, Hormonal control of metabolism in insects, *Adv. Insect Physiol.* **12:**270.
Steele, J.E., and Chan, F., 1980, Na^+-Dependent respiration in the insect nerve cord and its control by octopamine, in: *Insect Neurobiology and Pesticide Action* (M. Sherwood, ed.), p. 347, Society of Chemical Industry, London.

Stokes, T., and Awapara, J., 1968, Alanine and succinate as an end product of glucose degradation in the clam *Rangia cuniata, Comp. Biochem. Physiol.* **25**:883.

Strang, R.H.C., and Bachelard, H.S., 1973, Rates of cerebral glucose utilization in rats anaesthetized with phenobarbitone, *J. Neurochem.* **20**:987.

Strang, R.H.C., and Clement, E.M., 1980, The relative importance of glucose and trehalose in the metabolism of the nervous system of the locust *Schistocerca americana gregaria, Insect Biochem.* **10**:155.

Strang, R.H.C., Clement, E.M., and Rae, R.C., 1979, Some aspects of the carbohydrate metabolism of the thoracic ganglia of the locust *Schistocerca gregaria, Comp. Biochem. Physiol.* **62B**:217.

Sugden, P.H., and Newsholme, E.A., 1973, Activities of hexokinase, phosphofructokinase, 3-oxo-acid coenzyme A transferase and aceto-acetyl coenzyme A thiolase in nervous tissue from vertebrates and invertebrates, *Biochem. J.* **134**:97.

Sugden, P.H., and Newsholme, E.A., 1975, Activities of citrate synthetase, NAD^+-linked and $NADP^+$-linked isocitrate dehydrogenase, glutamate dehydrogenase, asparate amino-transferase and alanine amino-transferase in nervous tissues from vertebrates and invertebrates, *Biochem. J.* **150**:105.

Tietz, A., 1967, Fat transport in the locust: The role of diglyceride, *Eur. J. Biochem.* **2**:236.

Treherne, J.E., 1958, Facilitated diffusion and exchange absorption of glucose by the locust, *Schistocerca gregaria, Nature (London)* **181**:1280.

Treherne, J.E., 1960, The nutrition of the central nervous system of the cockroach *Periplaneta americana, J. Exp. Biol.* **37**:513.

Van der Horst, D.J., Van Doorn, J.M., and Beenakkers, A.M.T. 1978, Dynamics in the haemolymph trehalose pool during flight of the locust, *Locusta migratoria, Insect Biochem.* **8**:413.

Walker, P.R., and Bailey, E., 1969, A comparison of the properties of the phosphofructokinase of the fat body and flight muscle of the adult male desert locust, *Biochem. J.* **111**:365.

Wallace, B.C., Talamo, B.R., Evans, P.D., and Kravitz, E.A., 1974, Octopamine: Selective association with specific neurones in the lobster nervous system, *Brain Res.* **74**:349.

Weis-Fogh, T., 1952, Fat combustion and the metabolic rate of flying locusts, *Schistocerca gregaria* Forskål, *Philos. Trans. R. Soc. London, Ser. B* **237**:1.

Weis-Fogh, T., 1964, Diffusion in insect wing muscle, the most active tissue known, *J. Exp. Biol.* **41**:229.

Wigglesworth, V.B., 1960, The nutrition of the central nervous system in the cockroach *Periplaneta americana* L., *J. Exp. Biol.* **37**:500.

Williamson, D.H., Bates, M.W., Page, M.A., Krebs, H.A., 1971, Activities of enzyme involved in acetoacetate utilization in adult mammalian tissues, *Biochem. J.* **121**:41.

Worm, R.A.A., and Beenakkers, A.M.T., 1980, Regulation of substrate utilization in the flight muscle of the locust *Locusta migratoria* during flight, *Insect Biochem.* **10**:53.

8

Neuroendocrine Regulation of Mitochondrial Development and Function in the Insect Fat Body

L.L. Keeley

1. Introduction

The fat body is the principal insect tissue for intermediary metabolism. Through its production of hemolymph metabolites, the fat body influences the overall physiology of the animal. Factors that affect the biosynthetic capacity of this tissue secondarily influence most other physiological processes within the insect. The biosynthetic activity of the fat body cannot exceed the tissue's energy-generating capacity, which is determined by the number of functional mitochondrial units. Therefore, studies on the formation and function of fat body mitochondria are integral to an understanding of physiology and homeostasis in insects.

Little information exists concerning fat body mitochondria. The fat body is not a rich source of mitochondria, and reports on this subject are usually a minor aspect of a different topic. The present discussion will focus on the studies of my co-workers and me on the fat body mitochondria of the cockroach *Blaberus discoidalis*. Our findings are considered in the light of other information on fat body functions and along with results on mitochondriogenesis and its regulation in other organisms. The discussion will follow the historical development of the evidence for endocrine regulation of basal metabolic capacity in insects and will present

L.L. Keeley • Department of Entomology, Texas Agricultural Experiment Station, Texas A&M University, College Station, Texas.

our results on the role of neurohormones in fat body mitochondriogenesis and its current status.

2. Endocrine Regulation of Fat Body Respiration

2.1. Whole Body Respiration

Hormones influence basal metabolism in insects, as evidenced by respiratory changes after endocrine imbalance. Early studies focused on the role of the corpora allata (CA). For example, allatectomy (CA^-) decreased whole body respiration by 24% in adult, female *Calliphora erythrocephala* (Thomsen, 1949). Since the CA affect ovarian maturation in adult female insects (Wigglesworth, 1935; Pfeiffer, 1945), the depressive effects of CA^- on whole body respiration may not have been directly on basal metabolism but may have resulted from secondary responses to a reproductive disturbance. However, ovariectomy did not decrease the respiratory rate in adult *Calliphora* (Thomsen and Hamburger, 1955), which suggested that the earlier respiratory decreases following CA^- were not related to oocyte maturation but were direct metabolic responses to the absence of the CA hormone.

Later studies contradicted the previous results and failed to support a direct CA effect on basal metabolism in female insects. In adult *Leucophaea maderae*, respiratory cycles correlated with reproductive cycles (Sagesser, 1960), suggesting that the metabolic levels are a function of CA-directed, reproductive activities. Slama (1964) reported that CA^- lowered respiration in adult female *Pyrrhocoris apterus*, and cardiacectomy (CC^-), in addition to CA^-, depressed the O_2 uptake even further. He concluded that the CA affect reproduction-related metabolism and the corpora cardiaca (CC) affect basal metabolism. Since the CC are neurosecretory structures, Slama's results suggested that basal metabolism is neuroendocrine-regulated in insects.

My initial studies examined the influence of neurohormones on basal metabolism in adult male *Blaberus* cockroaches (Keeley and Friedman, 1967). The experiments determined the effects of chronic endocrine deficiency on whole body respiration rates. The use of males eliminated the cyclic nature of the metabolism that occurred in earlier studies with reproductive females. These studies demonstrated that 30 days after CC^- + CA^- the whole body O_2 uptake was reduced by 24% in adult male *Blaberus* (Keeley and Friedman, 1967). The respiratory levels were not affected significantly by either sham surgery or CA^-, when compared with unoperated animals. This indicated that chronic (30 days) endocrine deficiency following $CC^- + CA^-$ lowered the basal metabolic rate in adult male *Blaberus* and that the CC were clearly the effective glands.

The whole body respiration studies with *Blaberus* demonstrate that the neuroendocrine system can affect the basal metabolic capacity in insects. The previous test systems used by other workers obscured the neuroendocrine response by using females, in which the magnitude of the CA-directed, reproduction-related respiratory response was of the order of 70% to 100% (Slama, 1964). In contrast, the neuroendocrine response of basal metabolism is only 24%. The corpora allata may increase specific metabolic processes related to female reproduction, but the neuroendocrine system probably regulates the underlying basal metabolism, which determines the general biosynthetic capacity in the metabolic tissues. This agrees with the results of Slama (1964).

2.2. Tissue Respiration

Few studies of endocrine effects on whole body respiration have been carried to the tissue level. Slama (1965) examined the respiratory responses of adult body fragments to CA^- and diapause in female *Pyrrhocoris*. The thorax did not respond to the endocrine balance, indicating that, once formed, the thoracic musculature is functionally independent of hormones. However, abdominal respiration (gut, ovary, fat body) decreased after CA^- and during diapause, indicating that the metabolism of one or more of the abdominal tissues is hormone dependent. In adult *Leucophaea*, decapitation arrested ovarian development and lowered fat body respiration by 50% (Luscher, 1968). Implanted CA restored both the ovarian development and the fat body respiration rate. These results indicate that both the fat body and the ovary are targets for the CA hormone in female insects. No particular role for the CC was evident for the tissues of females in the experiments performed by Luscher (1968). In contrast, Muller and Engelmann (1968) found that chronic $CC^- + CA^-$ of female *Leucophaea* lowered fat body respiration more than did CA^-, and CC homogenates stimulated *in vitro* fat body respiration. The extent of the *in vitro* effect by the CC was dependent on the reproductive state of the fat body donor animal and was the inverse of the fat body's basal respiratory rate. Therefore, both the CC and the CA influence fat body metabolism in females, and the nature of the effect is complex and dependent upon the physiological state of the animal.

Respiration rates were measured for two major tissues in adult male *Blaberus* to determine their contributions to the CC-dependent respiratory decrease observed at the whole body level (Keeley and Friedman, 1967). The two tissues examined were the coxal muscles and the fat body.

Table 1. Effects of Chronic Cardiacectomy–Allatectomy on Oxidative Metabolism in 30-Day-Old Adult Male *Blaberus discoidalis*[a,b,c]

Tissue	Control[d]	$CC^- + CA^-$ [d]	P[e]
Coxal muscle	225 ± 12 (8)	219 ± 14 (9)	N.S.
Fat body:			
Intact	31 ± 4 (9)	17 ± 2 (9)	<0.01
Homogenate	29 ± 4 (4)	17 ± 1 (4)	<0.05

[a] Modified from Keeley and Friedman, 1967, 1969.
[b] Specific activity = nmol O_2 used/mg dry wt/hr.
[c] Intact tissue respiration was measured in a Warburg respirometer at 25°C in 3 ml of a medium containing 0.9% NaCl, 0.02% KCl, and 10 mM potassium phosphate (pH 7.0). Glucose (0.15 ml, 0.583 M) was tipped in from a side arm at time zero. Homogenate respiration was measured at 35°C by oxygen electrode polarography in a reaction medium containing 30 mM potassium phosphate, 15 mM KCl, 2 mM EDTA, 5 mM $MgCl_2$, 50 mM Tris(hydroxymethyl)aminomethane, 1 mM ADP, 30 mM glucose, and 20 mM ketoglutarate (pH 7.5). The test system consisted of 2.5 ml of the reaction medium and 0.5 ml of fat body homogenized in 154 mM KCl–1mM EDTA (pH 7.4): 1 fat body/ml.
[d] Values are mean ± S. E., with numbers of replicate tests shown in parenthesis.
[e] Statistical probabilities (P) are based on the Student's t test.

Chronic $CC^- + CA^-$ did not affect coxal muscle respiration, but fat body respiration was reduced by nearly 50% (Table 1). It was assumed that CC^- caused the reduced fat body respiration since it was responsible for the whole body effect. No other tissues were examined for their responses to $CC^- + CA^-$. Since the fat body of *Blaberus* is a large, metabolically significant tissue, it was used in further investigations on the cellular actions of the CC.

Previous investigators have described CC influences on fat body respiration. Additions of CC extracts directly stimulated *in vitro* O_2 consumption by fat bodies from *Leucophaea* (Luscher and Leuthold, 1965; Wiens and Gilbert, 1965; Muller and Engelmann, 1968). This direct stimulus to fat body O_2 uptake results from a shift in the principal substrates used for oxidative metabolism. Wiens and Gilbert (1965) reported that the addition of CC extracts to fat bodies *in vitro* increased O_2 uptake coincident with a drop in the respiratory quotient, thus indicating that the increased O_2 is used for the oxidation of lipids. The hypertrehalosemic effect of the CC increases the use of lipids as a fat body energy source while carbohydrates are diverted to trehalose synthesis (Wiens and Gilbert, 1965). These results demonstrate a direct effect by the CC on fat body respiration.

CC effects on fat body respiration were investigated in adult male *Blaberus* (Keeley and Friedman, 1969). Homogenates of fat bodies from $CC^- + CA^-$ animals respired with ketoglutarate at a 41% lower rate than homogenates of fat bodies from normal animals (Table 1). The continued low respiration in the presence of a Krebs cycle intermediate suggested that the enzymic lesion resulting from chronic CC deficiency is located in the electron transport chain between the Krebs cycle and O_2. This indicated that the CC deficiency affects fat body respiration at the mitochondrial level.

2.3. Fat Body Mitochondrial Respiration

Fat body mitochondria were isolated and characterized from adult male *Blaberus* (Keeley, 1971, 1973). The mitochondria were isolated by several procedures which gave identical results, but the preferred method was a rapid centrifugation procedure that took approximately ten minutes between the initial removal of the tissue and the final mitochondrial pellet (Keeley, 1973). Fat body mitochondria are similar to vertebrate liver mitochondria and are unlike the more intensively studied sarcosomes from insect flight muscles. Isolated fat body mitochondria are about 1 μm in diameter and have a condensed configuration. The final pellet is contaminated primarily by glycogen, which does not affect either the enzymic or protein analyses. Although the fat body of *Blaberus* has mycetocytes containing symbiotic bacteroids, the mitochondrial pellets are virtually devoid of bacteroid contaminants since these large organisms sediment with the cellular debris. Oxygen electrode polarography is used to monitor the O_2 uptake of the mitochondria, and the fat body mitochondria respire most actively with succinate. Ketoglutarate is used at an intermediate level of activity, but glycerophosphate, pyruvate–malate, and hydroxybutyrate are used only at low rates.

The respiratory and electron transport activities in fat body mitochondria from untreated 30-day-old adult male *Blaberus* were compared to those in similar $CC^- + CA^-$ animals (Keeley, 1971, 1972). In the following studies on chronic gland-deficiency effects, surgery was performed within five days of adult eclosion, and effects on fat body mitochondria were measured on 27- to 33-day-old adults. In all cases, the respiratory and electron transport activities were reduced by 40% in the fat body mitochondria from $CC^- + CA^-$ animals (Table 2). This confirmed our earlier speculation that the enzymic lesion resulting from chronic endocrine deficiency occurs in the region of the electron transfer enzymes. No single cytochrome appeared to be affected, since both the cytochrome c reductase and oxidase enzymes were low in activity. In contrast, the endocrine disruption did not change significantly the responsiveness to ADP (respiratory control) or the efficiency of oxidative phosphorylation (Table 2). The reduced respiration of the mitochondria suggests that the bioenergetic capacity of the fat body is 40% lower after

Table 2. Effects of Chronic Cardiacectomy–Allatectomy on the Respiratory Functions of Fat Body Mitochondria Isolated from 30 ± 3-Day-Old Adult Male *Blaberus discoidalis*[a,b]

Treatment	Q_{O_2}[c]	SCR[d]	COx[d]	RCR[e]	ADP/O
Unoperated[f]	4140 ± 170 (17)	154 ± 7 (17)	89 ± 6 (17)	3.3 ± 0.1 (17)	1.7 ± 0.1 (17)
CC^- + CA^-[f]	2540 ± 170 (13)	95 ± 9 (7)	54 ± 4 (8)	2.8 ± 0.2 (13)	1.5 ± 0.1 (13)
P[g]	<0.001	<0.001	<0.001		

[a] Modified from Keeley, 1972.
[b] Q_{O_2} was measured polarographically in 1.95 ml of reaction medium [30 mM potassium phosphate, 15 mM KCl, 2 mM EDTA, 5 mM $MgCl_2$, 50 mM Tris(hydroxymethyl)aminomethane, 2 mM ADP, and 20 mM sodium succinate (pH 7.4)] using 0.05 ml mitochondria (3–10 mg/ml) in 250 mM sucrose–2 mM EDTA (pH 7.4). RCR and ADP/O were measured in the same system containing 0.3% alcohol-extracted bovine serum albumin and 250 mM sucrose. Reductase and oxidase were measured as described previously (Keeley, 1972).
[c] Q_{O_2} = nmol O_2/mg protein/hr.
[d] SCR (succinate–cytochrome c reductase) and COx (cytochrome c oxidase) = nmol cytochrome c converted/mg protein/min.
[e] RCR = Respiratory control ratio = Q_{O_2} state 3 (+ADP) − Q_{O_2} state 4 (−ADP).
[f] Values are mean ± S.E., with numbers of replicate tests shown in parentheses.
[g] Statistical probabilities (*P*) are based on the Student's *t* test.

endocrine disruption, not because of a decreased efficiency for ATP synthesis, but because of a lower rate of ATP production by the decreased electron transport rate.

2.4. Effects of Specific Endocrine Glands

Specific surgical procedures and endocrine gland combinations were investigated for their effects on the respiration of fat body mitochondria in adult male *Blaberus* (Keeley, 1970, 1971). Sham surgery did not change mitochondrial O_2 uptake relative to unoperated animals (Table 3). CC^- + CA^- and CC^- (CA *in situ*) both depressed O_2 use significantly, but CA^- (CC *in situ*) had no effect on O_2 uptake. These findings clearly showed that the depressed respiratory activity in fat body mitochondria from CC^- + CA^- animals results from the absence of the CC.

Evidence that the active factor is of brain origin with subsequent transfer to the CC for storage and release was determined by severing the nervi corporis cardiaci I (NCC I) to the brain. The CC–CA complex was left *in situ* after severing the brain connectives. The status of the NCC II and III is unclear, since these nerves were not evident during surgery; however, the CC were displaced during the surgery to the extent that these other nervous connectives were probably broken, as well. Nerve severance reduced the rate of O_2 uptake to the level of CC^- + CA^- animals (Table 3). This finding suggests that the brain is the neurosecretory source of the respiration-stimulating factor found in the CC.

Hormone replacement studies confirmed that specific gland extracts

Table 3. Chronic Endocrine Deficiency Effects on the Oxidative Activity of Fat Body Mitochondria from 30 ± 3-Day-Old Adult Male *Blaberus discoidalis*[a,b]

Treatment	Qo_2[c,d]	P[e]
Untreated	3950 ± 190 (19)	—
Sham surgery	4440 ± 330 (9)	N.S.
CC^- + CA^-	3180 ± 260 (10)	<0.05
CA^-	4120 ± 240 (9)	N.S.
CC^-	3030 ± 100 (8)	<0.01
Severance of NCC I (CC + CA *in situ*)	3290 ± 230 (12)	<0.05

[a] Modified from Keeley and Waddill, 1971.
[b] Assay conditions are the same as in Table 2.
[c] Qo_2 = nmol O_2/mg protein/hr.
[d] Values are mean ± S.E., with numbers of replicate tests shown in parenthesis.
[e] Statistical probabilities (P) are based on Student's t test. Each experimental group was compared to the untreated control.

could reverse the CC^--related decrease in mitochondrial respiration (Keeley and Waddill, 1971). Extracts of specific endocrine glands, or brains, were injected daily between days 20 and 30 of adult age into chronic CC^- + CA^- animals. The total equivalent of one gland pair or brain complex was administered during the ten-day treatment period. Fat body mitochondria were isolated and respiration rates were measured immediately after the last extract injection.

The only active extracts were those prepared from the CC (Table 4).

Table 4. Effects of Endocrine Tissue Extracts on the Oxidative Activity of Fat Body Mitochondria from 30 ± 3-Day-Old Cardiacectomized–Allatectomized Adult Male *Blaberus discoidalis*[a,b]

Extract	Qo_2[c,d]	P[e]
Ringer	3040 ± 230 (6)	—
Brain + c. cardiaca	4190 ± 160 (6)	<0.01
Brain	3560 ± 130 (4)	N.S.
C. cardiaca	5270 ± 340 (6)	<0.01
Brain + c. cardiaca (boiled)	2570 ± 230 (5)	N.S.
C. allata	2950 ± 80 (4)	N.S.

[a] Modified from Keeley and Waddill, 1971.
[b] Assay conditions are the same as in Table 2.
[c] Qo_2 = nmol O_2/mg protein/hr.
[d] Values are mean ± S.E., with numbers of replicate tests shown in parenthesis.
[e] Statistical probabilities (P) are based on the Student's t test. Each group receiving an extract was compared statistically to the Ringer-injected controls.

Extracts of the brain + CC or the CC alone, resulted in mitochondrial respiratory rates that were equivalent to those found in normal animals. Brain or CA extracts had no significant stimulatory effects. The brain + CC extract lost its ability to increase mitochondrial respiration when boiled for 5 min. The heated brain + CC preparation also served as a tissue control for these studies and showed that nonspecific agents present in nervous tissues were not the cause of the respiratory stimulation by the CC extract.

Finally, the studies demonstrated that the CC acts indirectly on the respiration of the fat body mitochondria. Extracts of six CC pairs did not stimulate the respiration when added directly to fat body mitochondria isolated either from normal 30-day-old insects or from CC^- + CA^- 30-day-old insects (Keeley and Friedman, 1969). This indicated that the CC-extract injections stimulate the respiration of the fat body mitochondria indirectly through latent effects on the fat body adipocytes.

In summary, the results presented in Tables 2, 3, and 4 indicate a neuroendocrine regulation of fat body respiration in adult male *Blaberus*. Neurohormone deficiency caused by CC^- lowers the electron transport activity in the fat body mitochondria. Conversely, CC extracts stimulate mitochondrial respiration by acting through the adipocytes and not by direct activation of mitochondrial electron transport *per se*. Presumably, the hormone stimulates biosynthetic processes in the cell that contribute to the structural and/or enzymic integrity of the mitochondria and enhance their respiratory capacity.

3. Respiratory Development in the Fat Body

3.1. Patterns of Mitochondrial Respiratory Development

The results from the preceding experiments suggest that a putative neurohormone stored in the CC indirectly affects fat body respiration at the level of mitochondrial electron transport. Possibly, the neurohormone acts on processes related to the production and maintenance of new mitochondrial units in the fat body. Therefore, mitochondrial development was examined in the fat body of young adult male *Blaberus*.

Previous studies have described cycles of increase and regression for fat body mitochondria in correlation with insect growth and development. Large, swollen mitochondria were observed in the fat body of blowfly larvae during the early part of an instar (Marx, 1971; Tsuyama and Miura, 1979); however, later in the same instar, the mitochondria became structurally dense. These results indicate structural, and probably functional, changes in the mitochondrial population in response to the physiological state of the tissue. In the skipper butterfly, *Calpodes ethlius*, internal remodeling occurred in fat body adipocytes during metamorphosis (Larsen, 1976), and larval adipocytes were converted into adult adipocytes. Just before pupation, the metabolic organelles (mitochondria, microbodies, rough endoplasmic reticulum) were lost from the larval adipocytes, and protein, lipid, and RNA accumulated. In the adipocytes of young adults, the metabolic organelles were reformed and stored metabolites were lost. During the first 34 hr of adult life in *Calpodes*, fat body mitochondria increased in number by sevenfold (Larsen, 1970). Transverse inner membrane partitions were observed in the mitochondria, indicating that new mitochondria arose by growth and division of the existing population. The studies on *Calpodes* fat body demonstrate that the quality and quantity of the mitochondria are functions of the physiological state of the tissue, and new mitochondria are formed during early adult life.

Mitochondrial development during early adult life was evaluated in the fat body of *Blaberus* by measuring respiratory activities. The respiratory capacity was assessed at three levels: the intact fat body, isolated fat body mitochondria, and electron transport enzymes (Table 5). In all three cases, the activity increased by two- to threefold between days zero (adult eclosion) and ten of adult life. The mitochondria reached their maximal respiratory and electron transport activities between days six and eight, but tissue respiration did not reach a maximum until day ten. Since respiration reflects the total energy demands of the tissue, this latter

Table 5. Effects of Age and Endocrine Disruption on the Respiratory Activities of the Fat Body in Adult Male *Blaberus discoidalis*[a,b]

		Q_{O_2}[c,d]		Electron transport[c,d]		
Treatment	Age	Intact fat body	Mitochondria	Succinate–cytochrome c reductase	N	Cytochrome c oxidase
Untreated	0	17950 ± 420 (10)	1500 ± 230 (6)	50 ± 5	(10)	30 ± 3
”	2	18790 ± 1340 (8)	1860 ± 340 (6)	52 ± 4	(10)	34 ± 2
”	4	18300 ± 1000 (6)	2020 ± 170 (6)	80 ± 9	(7)	42 ± 5
”	6	22050 ± 1050 (8)	3800 ± 680 (6)	80 ± 5	(7)	42 ± 5
”	8	24460 ± 1010 (6)	3690 ± 430 (6)	140 ± 5	(7)	56 ± 4
”	10	31960 ± 1610 (9)	4240 ± 910 (6)	150 ± 5	(10)	59 ± 3
”	30	32320 ± 1040 (8)	4050 ± 640 (6)	150 ± 3	(10)	59 ± 2
$CC^- + CA^-$	30	24550 ± 1520 (8)	2490 ± 260 (6)	76 ± 4	(6)	37 ± 3

[a] Mitochondrial assay conditions are the same as in Table 2.
[b] Fat body respiration was measured by constant pressure manometry with individual fat bodies in 2 ml of Ephrussi–Beadle Ringer. All respiratory rates were measured at 35°C.
[c] Specific activities: Q_{O_2} intact fat body = nmol O_2 used/g dry wt/hr; Q_{O_2} mitochondria = nmol O_2 used/mg protein/hr; Reductase and oxidase = nmol cytochrome c converted/mg protein/min.
[d] Values are mean ± S.E., with numbers of replicate tests shown in parenthesis.

result indicates that the development of maximum respiratory capacity is completed in the mitochondria before the tissue increases its energy needs. No differences in oxidative activity were noted between days ten and thirty for any of the measured systems. Once the mitochondria completed their respiratory development by day six, they maintained this level for at least the next 20 days.

Measurements on respiratory control and ADP/O ratios assessed the energy-generating capacity of the mitochondria during the zero- to ten-day development period. The respiratory control and ADP/O ratios increased by only 30% between days zero and 30 (succinate substrate). (Keeley, 1972). In contrast, the rates of state 4* and state 3† respiration increased by 2.5 times and 3.5 times respectively, during the same period. The increases in respiratory control and ADP/O ratios were minor, relative to the increased electron transport capacity. This indicates that the major biochemical development in the fat body mitochondria is in the *rate* of energy generation and not in greater efficiency for energy conservation in high-energy bonds. This agrees with my earlier finding that chronic CC^- decreased electron transport but not the efficiency of ATP formation.

3.2. Endocrine Effects on Respiratory Development

Studies were made to determine if the previously identified neuroendocrine effects on the respiration of fat body mitochondria are related to respiratory development in young adult *Blaberus*. The respiratory levels of fat body mitochondria from 30-day-old, $CC^- + CA^-$ animals were compared with those of mitochondria from zero- to ten-day-old animals (Table 5). The respiratory levels of the mitochondria from the $CC^- + CA^-$ animals correspond generally to the four- to six-day levels of development. This finding suggests that endocrine deficiency prevents completion of respiratory development beyond its intermediate, five-day level.

Earlier studies (Keeley, 1971) had demonstrated that CA^- had no effect on mitochondrial respiration in *Blaberus*, but respiration was decreased by neuroendocrine disruption (CC^-). Therefore, the results presented in Table 5 suggest a neuroendocrine regulation over developmental events that occur around days four to six of adult life and that increase the respiratory capacity in the fat body mitochondria.

* State 4 respiration is respiration measured in the presence of substrate, oxygen, and phosphate, but lacking ADP as the acceptor for the high energy phosphate.

† State 3 respiration is respiration measured in the complete system including substrate, oxygen, phosphate, and ADP.

Table 6. Effects of Injections of Corpora Cardiaca Extracts on the Development of Respiratory Functions in Fat Body Mitochondria from 5-Day-Old Adult Male *Blaberus discoidalis*[a]

Treatment	Q_{O_2}[b,c,d]	*N*	*P*[e]
Untreated	2520 ± 400	6	N.S.
Ringer	2720 ± 270	4	—
Corpora cardiaca extract	4420 ± 180	5	<0.001

[a] Modified from Keeley, 1972.
[b] Q_{O_2} assay conditions were 1.9 ml of reaction medium [250 mM sucrose, 30 mM potassium phosphate, 15 mM KCl, 2 mM ETDA, 5 mM $MgCl_2$, 50 mM Tris (hydroxymethyl)aminomethane, 2 mM ADP, and 20 mM sodium succinate (pH 7.4)] and 0.1 ml mitrochondria (6–9 mg protein/ml). T = 35°C.
[c] Q_{O_2} = nmol O_2/mg protein hr.
[d] Values are mean ± S.E.
[e] Statistical probability (*P*) is based on the Student's *t* test. The Ringer was the basis for statistical comparison.

Since CC deficiency inhibits completion of respiratory development, CC-extract injection studies were conducted to measure the effects of hormone replacement on respiratory maturation (Keeley, 1972). Extracts of CC from five-day-old donor animals were injected daily into unoperated adult male *Blaberus* during their first five days of adult life. Control animals were either untreated or received Ringer solution instead of CC extract. Fat body mitochondria were isolated on day five and their oxidative activity measured. Mitochondria from five-day-old, Ringer-treated animals respired at the same partly developed rate as mitochondria from five-day-old, untreated animals (Table 6). However, mitochondria from five-day-old, CC-treated animals respired at rates comparable to those of the developed mitochondria isolated from fat bodies of ten to thirty-day-old animals. This finding indicated that CC factors stimulate precocious maturation of the respiratory capacity in fat body mitochondria. The respiratory maturation can occur, from the standpoint of the mitochondria, during the first five days of adult life if the hormone is present. The apparent absence of the hormone prevents the completion of respiratory maturation until day six. (This question is reexamined more critically in Section 6.)

4. Fat Body Maturation

The preceding studies indicate that the CC regulate the development of respiratory capacity in the fat body mitochondria of young adult *Blaberus*. In the absence of the CC, the respiratory activity fails to develop and remains indefinitely at 40%–50% below its maximum capacity. The physiological cause for mitochondrial maturation in the adult fat body is unclear, and the following studies investigated this question.

4.1. Fat Body Ultrastructure

The fine structure of the fat body was examined during the respiratory maturation period of young adult male *Blaberus*. The cytoplasm in adipocytes of zero-day animals was found to be divided into storage and metabolic regions (Fig. 1). The storage regions were electron-lucent, lacked abundant metabolic organelles (ribosomes, mitochondria) and contained glycogen and lipid stores. The electron-dense, metabolic cytoplasm was distributed peripherally along the plamsa membrane and contained occasional, indistinct, ovoid mitochondria along with ribosomes and rough endoplasmic reticulum. At this same time, "protein storage bodies" were forming. The protein bodies were comprised of membranes and mitochondria ingested from the cytoplasm (Fig. 2). These bodies were, in fact, cytolosomes that digest cytoplasmic organelles and are indicative of cytoplasmic deterioration (Locke and Collins, 1965).

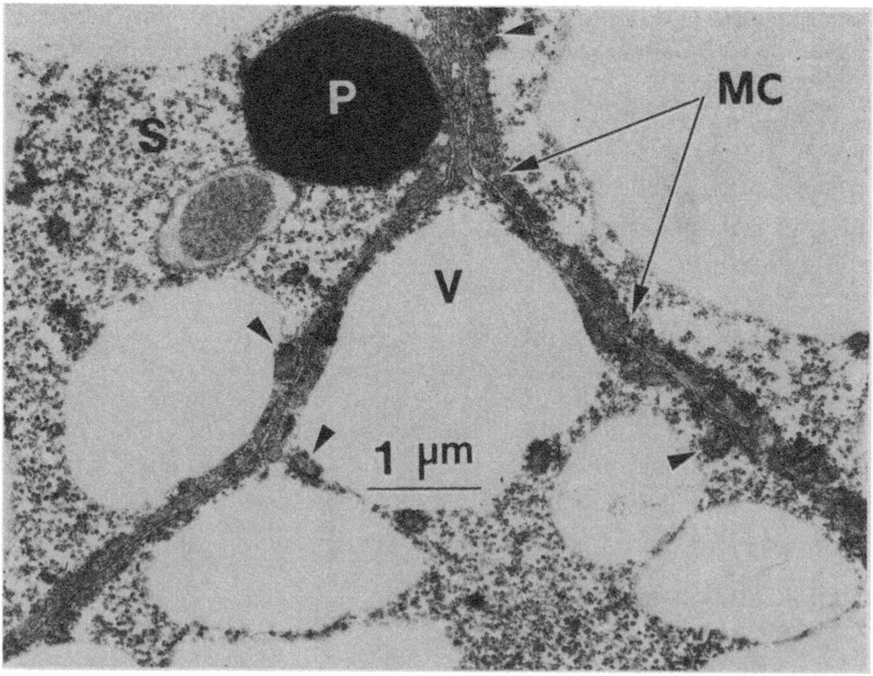

Fig. 1. Fine structure of an adipocyte from zero-day-old adult male *B. discoidalis*. Arrows indicate mitochondria; MC = metabolic cytoplasm; S = storage cytoplasm; P = protein body (cytolosome?); V = lipid vacuole.

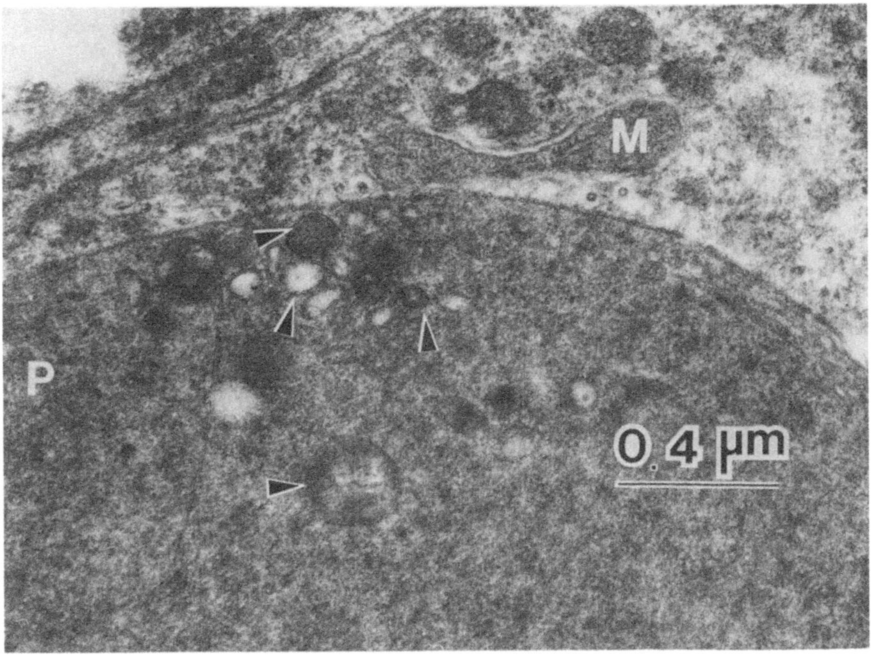

Fig. 2. Cytolosome (protein body) in an adipocyte of one-day-old adult male *B. discoidalis*. Note the mitochondria-like structures (arrows) and cytoplasmic membranes located along the periphery of the protein body (P). A normal mitochondrion (M) is evident in the adjacent cytoplasm.

Structural changes intensified during days one, two, and three of adult life (Fig. 3). Light microscopy revealed frequent coalescence between lipid vacuoles, suggesting lipid mobilization (Thomsen and Thomsen, 1974). Small ellipsoid mitochondria lacking structural definition were found frequently, and larger, elongate mitochondria were observed occasionally with some evidence of internal cristae (Figs. 2, 3). Cytolosomes were still prevalent during this time and suggested that the fat body cells were continuing their cytoplasmic breakdown. Ribosomes or rough endoplasmic reticulum were observed in the perinuclear area and in the metabolic cytoplasm along the cell membranes.

The adipocytic cytoplasm began to reestablish its discrete structure on day three and was completed by day four (Fig. 4). Storage and metabolic cytoplasms were no longer distinguished. By day four, well defined, elongate, densely staining mitochondria were numerous, and cytolosomes had become infrequent. Rough endoplasmic reticulum and ribosomes were abundant throughout the cytoplasm in general. Older adults (30 days) have abundant, intensely stained mitochondria throughout the cytoplasm of the adipocytes.

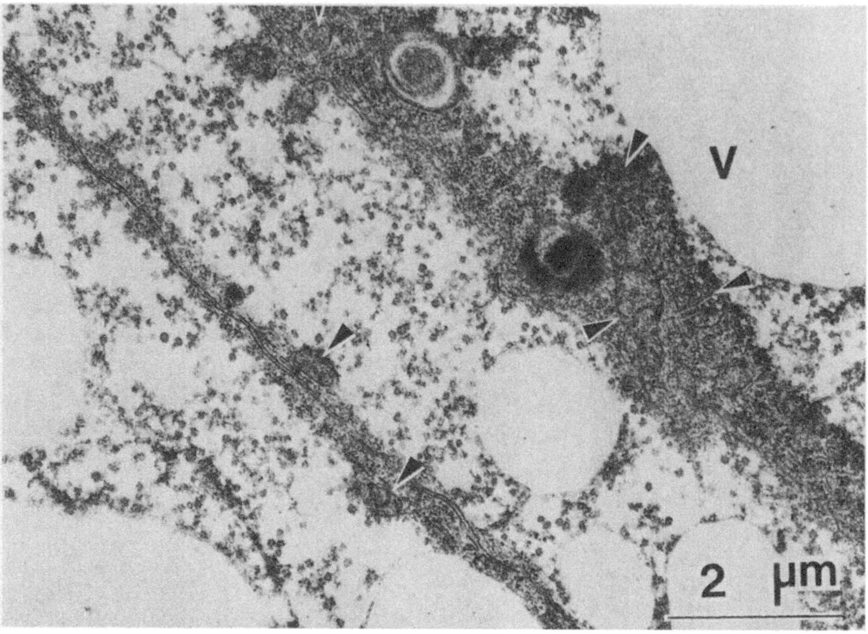

Fig. 3. Fine structure of an adipocyte from two-day-old adult male *B. discoidalis*. Mitochondria are indicated by arrows in the metabolic cytoplasm. V = lipid vacuole.

4.2. Changes in Metabolite Contents

The fine structure studies suggested that the fat body undergoes a structural rearrangement during days one, two, and three of adult life. Histochemical studies suggest that the contents of stored metabolites change during the same ages (Keeley and Thurston, unpublished observations). To confirm these changes, the major storage metabolites and RNA content of the fat body were quantitated during the early development period in adult male *Blaberus* (Mannix and Keeley, 1980).

Dry weight, glycogen, lipid, and protein displayed similar patterns of change in the fat body during the first ten days of adult life (Mannix and Keeley, 1980). The dry weight and the content of all three metabolites decreased by days two to three from their zero-day levels, increased on day four, then declined until they reached a level on day eight, at which they remained through day 10 (Fig. 5). Unlike glycogen, lipid, and protein, the RNA content of the fat body increased continuously between days zero and five, then remained constant at twice its initial level throughout the remainder of the 10-day study period. It was found that the metabolite levels present on day 10 remained constant through day 30 (Mannix and Keeley, 1980).

The fine structure observations and the changes in the metabolite contents provide complementary evidence that the adipocytes of *Blaberus* undergo a structural–biochemical transition during the first four days of adult life. The DNA content of the fat body does not change in *Blaberus* during its adult life (Mannix and Keeley, 1980), indicating that the number of fat body cells remains constant. Therefore, the adipocyte transition is probably an intracellular, postimaginal metamorphosis that converts a nymphal tissue into an adult tissue. The peak transition occurs on days two to three, based on the loss of metabolite contents and the organelles present in the cytoplasm. The initial, zero- to one-day fat body is apparently a carry-over of nymphal tissue used primarily for nutrient storage. Around days two and three, the adipocytes undergo metamorphosis by cytoplasmic reorganization and form an adult fat body that has a greater metabolic capability for biosynthetic functions. By day four, the cytoplasmic transition is complete, and the formation of new mitochondrial units increases the energy-generating capability of the cells to meet the biosynthetic demands of adult life. The neuroendocrine stimulus to mitochondrial development is part of the increase in energy-generating ca-

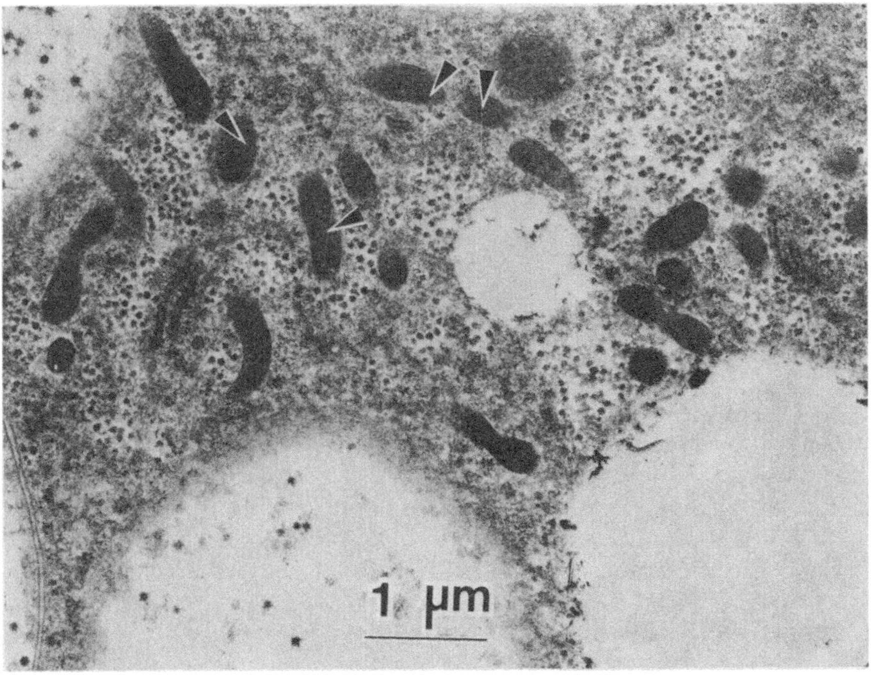

Fig. 4. Fine structure of an adipocyte from four-day-old adult male *B. discoidalis*. Note the general distribution of the metabolic cytoplasm and the abundant, distinct mitochondria (arrows).

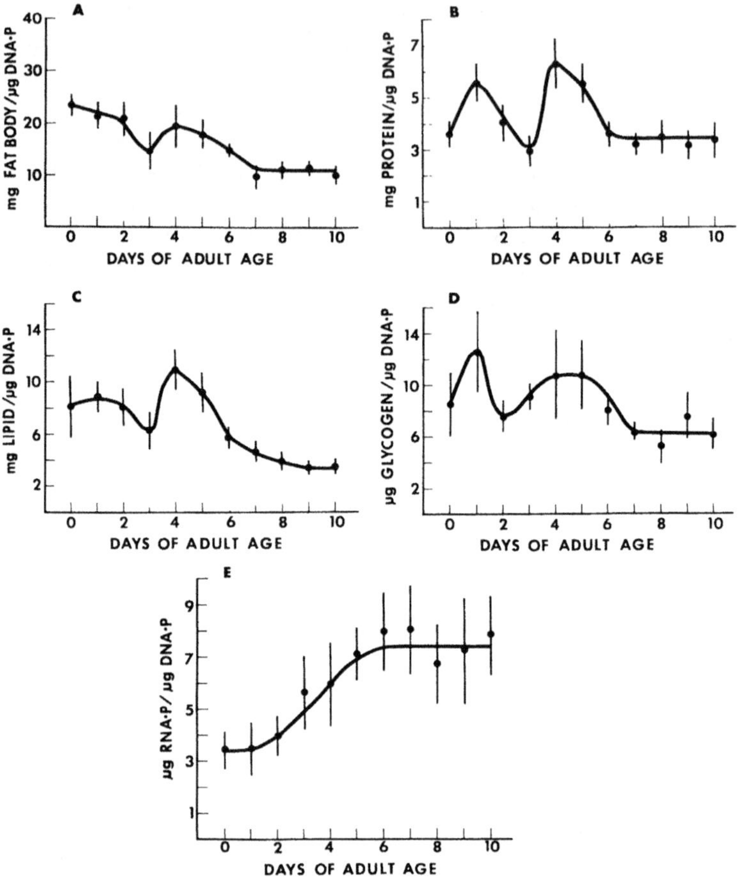

Fig. 5. Age-related changes in the metabolite contents of the fat body of adult male *Blaberus discoidalis* (from Mannix and Keeley, 1980). Values are means, with ±S.E. denoted by vertical bars. Each point represents the mean of six to ten replicates of two animals per replicate. DNA · P = quantity of fat body DNA as based on its phosphate content using a herring sperm DNA (P = 6.8%) as the standard; RNA · P = the quantity of fat body RNA as based on its phosphate content using a yeast RNA (P = 8.5%) as the standard.

pacity. Alternatively, we cannot ignore the possibility that structural reorganization and mitochondriogenesis may be growth-related processes that occur at each molt in conjunction with fat body enlargement.

5. Endocrine Regulation of Mitochondrial Development

5.1. Cytochrome Content

The uncoupling agent 2,4-dinitrophenol (2,4-DNP) is used to identify the rate-limiting processes of electron transport. Oxidation is uncoupled

Table 7. Effects of 2,4-Dinitrophenol on the Respiration Rate of Fat Body Mitochondria from Normal and Corpora Cardiaca–Allata Deficient 30-Day-Old Adult Male *Blaberus discoidalis*[a,b]

	Q_{O_2} coupled[c,d]		Q_{O_2} uncoupled (+DNP)[c,d]		
Experimental animals	State 4	State 3	State 4	State 3	*N*
Normal	1300 ± 130	4290 ± 270	3700 ± 310	3800 ± 180	7
$CC^- + CA^-$	1030 ± 130	2500 ± 360	2140 ± 310	2010 ± 310	5
	(N.S.)	(P < 0.01)	(P < 0.01)	(P < 0.01)	

[a] Modified from Keeley, 1972.
[b] Measurements of respiratory rates were performed using 1.8 ml of the reaction medium described in Table 6 and 0.2 ml of mitochondria (0.7–2.0 mg protein); 0.01 ml of 10 mM 2,4-DNP was added for uncoupling.
[c] Q_{O_2} = nmol O_2/mg protein/hr.
[d] Values are mean ± S.E. Probability is based on the Student's *t* test.

from phosphorylation by 2,4-DNP. Uncoupling relieves respiratory control by ADP and allows electron transport to "run free" at its maximum rate. In my studies, 30-day-old untreated or $CC^- + CA^-$ animals were compared for their responses to 2,4-DNP (Keeley, 1972). After measuring the initial oxidation rates in states 3 and 4, a second aliquot of mitochondria was placed in state 4, and 2,4-DNP (final conc. = 50 μM) was added. Uncoupled state 4 rates matched, but did not surpass, the coupled state 3 rates (Table 7). This held true for mitochondria from both untreated and $CC^- + CA^-$ animals. The addition of ADP to uncoupled mitochondria has no stimulatory effect. This indicates that respiration is rate-limited by the electron transport capacity in the fat body mitochondria. Since CC deficient animals have reduced electron transport capacities, the CC appear to effect the cytochrome content in the mitochondria.

The cytochrome content of the fat body mitochondria was measured during the respiratory development period (Table 8). The level of cyto-

Table 8. Cytochrome Complement in the Fat Body Mitochondria of Adult Male *B. discoidalis* during Respiratory Development

Adult age		Cytochrome content[a,b,c]			
(days)	*N*	c	c_1	b	aa_3
0	9	134 ± 11	199 ± 20	195 ± 25	157 ± 39
4	8	158 ± 8	192 ± 18	253 ± 32	173 ± 26
6	10	218 ± 18	225 ± 15	305 ± 30	275 ± 28

[a] Values are mean ± S.E. *N*: number of replicates of four animals per replicate.
[b] Cytochrome content: pmol/mg protein.
[c] Cytochromes were estimated by the method of Williams (1964).

chromes aa_3, b, and c increased during the six-day development period. Cytochrome c_1 was the only exception, and its level did not change during the respiratory development. The content of cytochromes aa_3 and b doubled, whereas that of cytochrome c increased by 48%.

The increasing cytochrome content indicates mitochondrial cytochrome synthesis is active in the fat body during the first six days of adult life and accounts for the respiratory development. In the absence of the CC, respiratory development stops at an intermediate level (Table 5), presumably as a result of reduced cytochrome synthesis. Therefore, neuroendocrine effects on mitochondrial cytochrome synthesis were examined in the fat body of *Blaberus*.

5.2. Cytochrome Synthesis

Cytochromes consist of a heme moiety and an apoprotein comprised of up to seven subunits. The apoproteins and the hemes show mutual feedback interactions on their synthesis and assembly into holocytochromes.

Regulation of heme biosynthesis is well known. The regulatory enzyme for heme formation is the mitochondrial enzyme aminolevulinic acid synthase (ALAS), which catalyzes the production of aminolevulinic acid (ALA) from glycine and succinate (Granick and Urata, 1963). Once formed, ALA is converted rapidly to heme. Vertebrate liver ALAS is an inducible enzyme with a half-life of one to two hours (Beattie and Stuchell, 1970). Steroid hormones induce ALAS activity in the mitochondria of avian liver cell cultures (Granick and Kappas, 1967), which suggests that ALAS levels may be hormonally regulated *in vivo*, as well. Finally, ALAS activity is end product inhibited by hemin (Scholnick *et al.*, 1972). These properties suggest that ALAS is a potential regulator for fat body cytochrome synthesis and may be sensitive to hormone regulations.

Heme synthesis may regulate cytochrome synthesis. For example, all cytochromes were absent in ALAS-deficient yeast mutants but reappeared after the addition of ALA to the culture medium (Woods, *et al.*, 1975). Saltzgaber-Muller and Schatz (1978) found that subunits I, IV, V, and VII of the apoprotein of cytochrome oxidase were barely detectable in the ALAS-deficient yeast mutants, while subunits II, III, and IV were present but unassembled. The addition of ALA resulted in fully formed cytochrome oxidase, and this suggested that the appearance of the heme induces both the formation of the missing apoprotein subunits and their assembly into the holocytochrome. In rat liver mitochondria, the induction of ALAS activity raised the cytochrome content and the *in vitro* incorporation of amino acids by 30% to 50% (Beattie and Stuchell, 1970;

Beattie, 1971). The latter results suggest that ALAS induction may activate several biosynthetic events related to mitochondriogenesis.

The relationship of ALAS and ALA to mitochondrial cytochrome synthesis in insects has been examined only with respect to flight muscle formation during metamorphosis. Injections of ALA into developing pupae of polyphemus moths increased the thoracic cytochrome c content by 68% when compared with untreated insects (Soslau *et al.*, 1971). In contrast, injections of ALA into developing *Manduca sexta* pupae did not stimulate cytochrome c synthesis, and no ALAS activity was measurable before the 17th day of development (Chan *et al.*, 1973). Four days later, at adult emergence, the ALAS activity of *Manduca* increased rapidly to a peak and then declined to an undetectable level within 18 hr. In a series of papers on the honeybee, it was reported that the level of cytochrome c began increasing in the developing flight muscles 24 to 48 hr before adult emergence and had increased by 40-fold by the time of emergence (Osanai and Rembold, 1970). The ALAS activity and ALA content increased steeply, simultaneously with the appearance of cytochrome c and reached their maximum levels in the newly emerged adult honeybee (Osanai and Rembold, 1974, 1975). During pupal metamorphosis of the bollworm *Heliothis zea*, fat body ALAS activity increased by nearly 100-fold five days before adult emergence (Gruetzmacher and Keeley, 1981), but injections of [^{14}C]-ALA did not stimulate tissue cytochrome synthesis for fat body mitochondria until immediately before emergence.

There are conflicting reports about the ability of administered ALA to stimulate cytochrome synthesis *in vivo*, but it is feasible that none of these reports are incorrect. ALA injections may increase cytochrome levels in tissues when the cytochrome synthetic capacity is present but operating at suboptimal levels owing to restricted ALA pools. However, ALA, by itself, probably cannot induce the capacity for cytochrome synthesis, and, in the absence of active cytochrome synthesis, exogenous ALA has no effect.

The capacity for mitochondrial cytochrome synthesis was measured in the fat body of *Blaberus* during the respiratory development period. Cytochrome synthesis was assessed by measuring ALAS activity and by the rate of incorporation of [^{14}C]-ALA into cytohemes.

5.2.1. Aminolevulinic Acid Synthase Activity

The possible regulatory role of ALAS for cytochrome synthesis was measured in the fat body mitochondria of *Blaberus* (Keeley, 1978a). ALAS activity showed a developmentally related pattern, with a maxi-

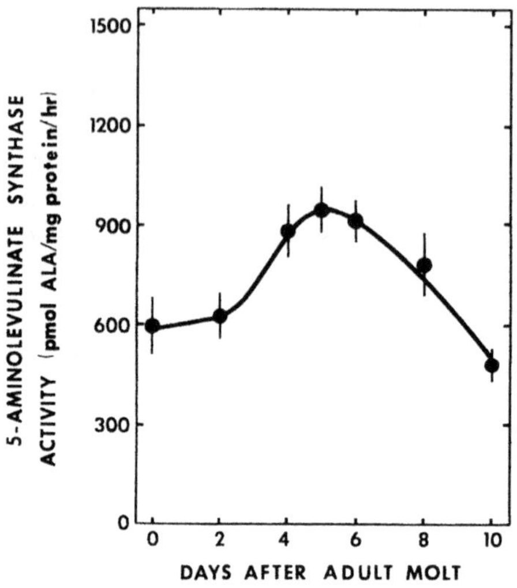

Fig. 6. Age-related aminolevulinic acid synthase activity in fat body mitochondria of adult male *Blaberus discoidalis*. Values are the mean of 8–10 individuals, with ± S.E. denoted by the vertical bars. ALAS activity was measured at 25°C. (Reprinted with permission, Keeley, 1978a.)

mum on days four to six of adult life (Fig. 6). The peak of activity on day five was 60% above the initial zero-day level. Fat body ALAS was induced by 245% above the control level by allylisopropylacetamide and had a half-life of 6 hr (Keeley, 1978a). If one accounts for the differences in body temperature between rats (37°C) and *Blaberus* (25°C), and assuming a Q_{10} effect, then the turnover rate of ALAS is similar in both the rat liver and the cockroach fat body.

Studies on endocrine regulation of ALAS activity were inconclusive. Results from chronic gland deficiency studies suggested that CA^- reduces ALAS activity relative to sham surgery controls (Keeley, 1978a). Conversely, insects treated with juvenile hormone-I (JH-I) increased their ALAS activity twofold, but olive oil-treated controls also increased their activity and were not statistically different from the JH-treated group. The data for JH regulation of ALAS activity in insects remains inconclusive. Neither CC extracts nor β-ecdysone increased ALAS activity. Therefore, although ALAS did show properties similar to those of the rat liver enzyme, it did not show characteristic responses to endocrine balance that would indicate it to be the endocrine-regulated step of fat body cytochrome synthesis.

5.2.2. [^{14}C]Aminolevulinic Acid Incorporation

[^{14}C]-ALA incorporation into cytohemes measures the capacity for fat body cytochrome synthesis independently of ALAS activity (Keeley, 1978b). It was proposed above that exogenously administered ALA incorporated into cytochromes only in proportion to the cell's endogenous capacity for cytochrome synthesis. This is the basis for the following experiments. In vertebrate animals, [^{14}C]-ALA is a specific label for heme and is not reused. We suspect the same is true of insects, unless an undescribed biochemical pathway is operative. The specificity for measuring cytochrome synthesis was increased in our studies by separating cytohemes a + b by acidic acetone extraction from cytochromes c + c_1 (Basford *et al.*, 1957).

The *in vivo* cytochrome synthetic capacity was measured for the fat body mitochondria of *Blaberus*. Animals were treated with [^{14}C]-ALA, and held for 4 hr at 25°C; the mitochondria were isolated, and cytohemes a + b were separated from cytochromes c + c_1 (Keeley, 1978b). Studies during the mitochondrial development period showed a peak of cytochrome synthesis on days four to six of adult life (Fig. 7). The peak activity was particularly dramatic for the synthesis of cytochromes aa_3 + b, which was nearly twice its zero-day level.

The age of maximum [^{14}C]-ALA incorporation coincided with the age of maximum ALAS activity (Fig. 6) and indicated that four to six days of adult age is the time of greatest cytochrome synthesis for the fat body mitochondria. The large increase in the synthesis of cytochromes aa_3 + b agreed with the increase in the mitochondrial content of these cytochromes during this same time (Table 8).

The earlier respiration studies had suggested that the neuroendocrine-regulated developmental (biosynthetic?) events occur in the fat body mitochondria around days four to six of adult life (Keeley, 1972). The peak of [^{14}C]-ALA incorporation into cytochromes around this same age suggests that the endocrine effect may be to stimulate cytochrome syn-

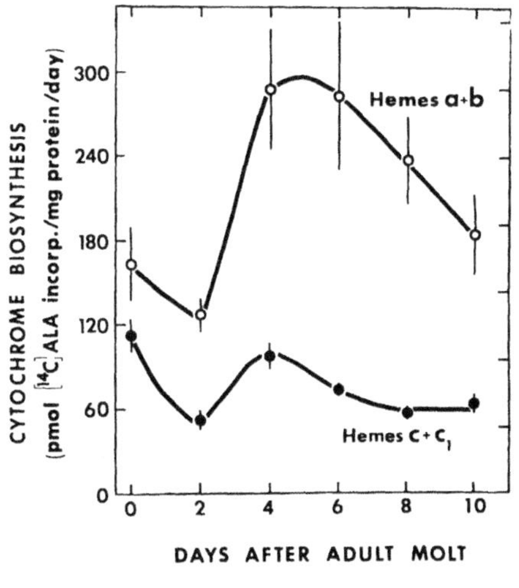

Fig. 7. Age-related [^{14}C]aminolevulinic acid incorporation into cytochromes of fat body mitochondria in adult male *Blaberus discoidalis*. Each point represents the mean ± S.E. of 8–9 individuals. Animals of the appropriate age were injected with 1 μci of [^{14}C]-ALA (= 19 nmol, 53 mCi/mmol) and incubated for 4 hr at 25°C before isolation of the hemes. (Reprinted with permission, Keeley, 1978b.)

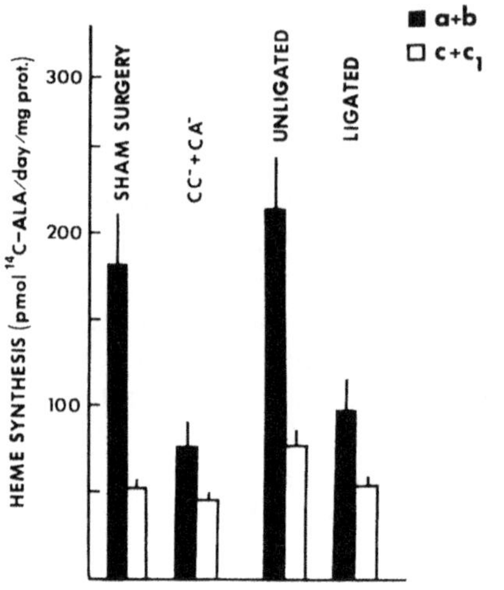

Fig. 8. Effects of cardiacectomy–allatectomy and ligation on the rate of *in vivo* [^{14}C]aminolevulinate incorporation into cytochrome hemes of fat body mitochondria in four-day-old adult male *Blaberus discoidalis*. Values are means, with + S.E. indicated by the centered vertical bars for six to nine individuals per test group. Animals were administered 0.5 μCi [^{14}C]-ALA (= 19 nmol ALA; 26.5mCi/mmol) and were incubated for 4 hr at 25°C before isolation of the hemes.

thesis. Since ALAS does not respond to CC^-, it is felt that the neuroendocrine-regulated events of cytochrome formation are related to either the conversion of ALA into heme or the assembly of heme and its apoproteins. Both of these would be reflected in our measurements on the rates of [^{14}C]-ALA incorporation into cytochrome hemes.

Endocrine effects on cytochrome synthesis were tested by extirpation of the CC and CA from newly emerged adult male *Blaberus*. Control animals were sham operated to ensure comparable stress effects on feeding and other processes. In a parallel study, zero-day animals were cervically ligated to eliminate the endogenous hormone secreted from the head. The four-day peak of [^{14}C]-ALA incorporation into cytochromes aa_3 + b was eliminated by both methods of rendering the animals endocrine deficient (Fig. 8). The synthesis of cytochromes aa_3 + b was reduced by about 55% after either form of endocrine deficiency. Cytochromes c + c_1 were unaffected by CC^- + CA^- but were reduced by ligation.

CC extracts were tested for their capacity to stimulate [^{14}C]-ALA incorporation into the cytochrome hemes. Animals ligated on day zero were used for these studies to eliminate any effects by endogenous hormones or nutrition. In keeping with the earlier studies showing positive respiratory responses from a regimen of daily injections, CC extracts (0.33 CC/day) were administered on days two, three, and four of adult life (48, 24, and 0 hr preassay). Immediately after the last injection on day four (= 0 hr preassay) [^{14}C]-ALA was administered and the label incorporation rate was measured as described previously. Gland extracts were prepared from the CC of one-day-old donors because any neurohormone regulating respiratory capacity should be especially abundant before the onset of the respiratory development.

The CC extracts stimulated [^{14}C]-ALA incorporation into the structural cytochromes aa_3 and b (Keeley, 1978b). Ligation reduced the syntheses of cytochromes aa_3 + b by 64% and of cytochromes c + c_1 by 43% (Fig. 9). The injections of CC extracts returned the synthetic capacity for cytochromes aa_3 + b to the level of normal four-day-old, untreated insects, but cytochromes c + c_1 did not respond. Injections of Ringer solution or extracts of muscle, fat body, or thoracic ganglia failed to stimulate the cytochrome synthesis. These results demonstrate that the increased synthesis of cytochromes aa_3 + b is a specific response to the CC extract and not a general response to injury or tissue extracts. Furthermore, the responsiveness of the ligated animals to the CC extract indicates that the CC are capable of causing the entire effect, independent of nutrition or other hormones secreted from the head.

Administrations of β-ecdysone, JH-I and JH-III, proctolin, and triidothyronine and thyroxine all failed to stimulate the synthesis of cytochromes aa_3 + b.

Although the synthesis of the c-type cytochromes was reduced by ligation and decapitation, these cytochromes did not respond to either CC extirpation or CC extract treatments. Instead, the decreased synthesis of the c-type cytochromes after ligation or decapitation may result from nutritional deficiency. Therefore, the results indicate that the neuroendocrine stimulus specifically affects the synthesis of cytochromes aa_3 and/or b during the respiratory development of the fat body mitochondria.

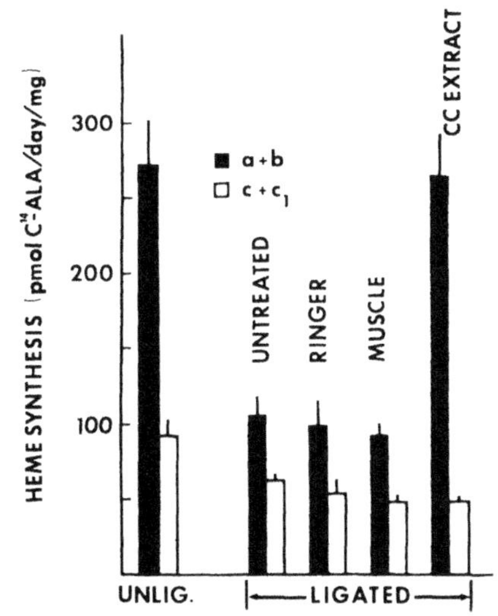

Fig. 9. Effects of corpora cardiaca extract injections on the *in vivo* incorporation of [^{14}C] aminolevulinate into the cytochrome hemes of fat body mitochondria in ligated, four-day-old adult male *Blaberus discoidalis*. Values are the means, with +S.E. indicated by the centered vertical bars for seven to fourteen individuals per test group.

Animals were ligated on day zero and injected daily with 10 μl of the appropriate extract (one CC/30 μl) or Ringer solution starting on day two of adult life. After the last treatment on day four, all test and control animals were administered 0.5 μCi of [^{14}C]-ALA (19 nmol; 26.5 mCi/mmol) and incubated for 4 hr at 25°C before isolation of the hemes.

6. Cytochromogenic Factor

My co-workers and I have initiated studies on the nature of the neurohormone that affects the synthesis of cytochromes aa_3 + b in the fat body of adult *Blaberus*. We are designating this agent as the "cytochromogenic factor" (CGF), and we have established a simple, reasonably rapid bioassay for CGF activity. The bioassay is based on stimulating cytochrome synthesis in decapitated animals.

In our early studies, CC extracts were always administered as a series of daily injections. The need for a regimen of daily CC extract treatments was determined by administering all sequential combinations of the CC extract injections (Hayes and Keeley, 1981). Regardless of the injection series, each test animal received the total equivalent of one pair of CC. In all cases, the CC extract regimen increased the synthesis of cytochromes aa_3 + b above that of comparably injected muscle extract controls. Those test groups receiving CC extract in combinations with a 48-hr preassay series (48 + 24 + 0 hr; 48 + 24 hr; and 48 hr) always had the greatest activities. The largest response was found in the three injection (48 + 24 + 0 hr preassay) series; the least response was found in the group that received CC extract and [^{14}C]-ALA simultaneously (= 0 hr preassay). These results indicate that CGF requires a 24 to 48 hr latent period, presumably for the synthesis of enzymes or components involved in the production of cytochromes aa_3 + b.

In a previous study, we speculated that the absence of CGF early in adult life prevents the respiratory maturation from occurring before day 6 (Section 3.2). The studies presented here contradict this idea by showing that CGF must, in fact, be present by day two, i.e., 48 hr before the peak cytochrome synthesis on day four. The maximum respiratory activity is reached around days six to eight because the CGF-induced peak of cytochrome synthesis starting on day four completes the cytochrome complement by day six. In the previous study, precocious respiration rates were evident in five-day-old adults treated from day zero with CC extracts. This presumably resulted from advancing the time of maximum cytochrome synthesis from day four to day two by the early presence of CGF. Preliminary studies on age-related CGF titers in the CC confirm the loss of CGF from the glands on days two to four, presumably due to secretion (Hayes and Keeley, unpublished results). Furthermore, decapitation before day two blocked the four-day increase in the synthesis of cytochromes aa_3 + b, but decapitation after day two did not block the four-day peak of cytochrome synthesis. These data indicate the secretion of CGF occurs between days two and three.

A dose-response was confirmed between the CC extracts and the synthesis of cytochromes aa_3 + b (Hayes and Keeley, 1981). The dose-response was linear between 0.01 and 0.075 CC pairs. All treatments gave a maximum response with more than 0.075 CC pairs total dosage.

The properties of CGF indicate that it is a small protein. It is sensitive to chymotrypsin, aminopeptidase, and carboxypeptidase, but insensitive to trypsin. It is dialyzable through a system that retained > 12,000 mol. wt. It is also sensitive to heat (100°C, 15 min). No part of the nervous system other than the CC has a significant amount of CGF activity. We suspect that CGF is the same factor described earlier (Section 2.4) and is probably synthesized by the brain and secreted from the CC.

These properties suggest, at present, that CGF is a newly discovered insect neurohormone. The properties of CGF do not correspond to any of the currently described neurohormones (Frontali and Gainer, 1977).

7. Significance and Conclusions

Few previous studies in insects have related hormones to mitochondriogenesis or to mitochondrial functions. Most such studies concerned the reformation of mitochondria following endocrine stimuli for diapause termination (see review: Shappirio, 1960); however, in this case the mitochondria respond to the hormones secondarily, following the resumption of growth or development. The same appears to be true for the mitochondriogenesis that occurs when flight muscles form during metamorphosis. Once the hormones stimulate the myoblasts to differentiate into muscles, mitochondrial formation accompanies muscle formation independently of further hormonal stimuli. Only Chaudhary and Malhotra (1974) have suggested that there are direct endocrine influences on muscle mitochondriogenesis. They reported an ecdysone-related increase in amino acid incorporation into flight muscle sarcosomes of *Schistocerca gregaria*.

Most of the studies of endocrine effects on insect mitochondria have focused on JH because of its effects on respiration. However, these studies have provided conflicting data, and no conclusions can be reached. In an early study, Clarke and Baldwin (1960) reported that crushed CA injected into adult locusts increased the respiratory activity of both the animals and their isolated mitochondria. In contrast, Minks (1967) reported that the addition of crushed CA or JH-active cecropia oil to isolated locust mitochondria increased P/O ratios but not respiration rates. However, both of these studies with CA homogenates are questionable. JH is secreted as soon as it is formed by the CA (Pratt *et al.*, 1975); therefore, it is unlikely that the addition of a CA homogenate is useful as a source of JH. In contrast to Minks, Chefurka (1978) reported that natural sesquiterpenes with JH activity acted as uncouplers for oxidative phosphorylation. In the last larval instars of *Dermestes maculatus*, JH analogs directly stimulated *in vivo* respiratory rates, but this effect was diet related (Slama and Kryspin-Sorensen, 1979). The JH stimulated extensive fatty acid metabolism, and the stimulus was lost if the insect was fed a lipid-free diet. Finally, Firstenberg and Silhacek (1973) made the interesting observation that the direct addition of JH to *in vitro* mitochondria suppressed the oxidation of NAD-dependent substrates but stimulated succinate oxidation. The physiological significance of this finding is unclear, and it is not known if it occurs *in vivo*.

These reports show no consistent evidence that JH has a direct physiological role on either mitochondrial functions or integrity. Instead, the data presented in this review indicate a physiological role for a neurohormone that regulates the synthesis of particular cytochromes during naturally occurring mitochondriogenesis in the fat body of young adult insects.

Our research indicates that the fat body of *Blaberus* is useful as a model for studies on the regulation of mitochondriogenesis in higher animals. Unlike some insect species in which fat body metamorphosis occurs by cytolysis and regrowth, the fat body of *Blaberus* matures after adult eclosion by cytoplasmic reorganization within existing adipocytes. Comparable adipocyte metamorphosis also occurs in newly formed adult *Calpodes* butterflies (Larsen, 1976) and in mosquitoes (Clements, 1963). Presumably, imaginal metamorphosis of adipocytes makes the fat body competent for adult biosynthetic functions, especially reproduction. Alternatively, the adipocyte changes may be growth related and occur generally after a molt. In any case, mitochondriogenesis is a facet of the fat body reorganization. As described here, fat body mitochondriogenesis is rapid, synchronized, and easily timed from adult eclosion. Certain biosynthetic aspects of the mitochondriogenesis are endocrine regulated, especially those dealing with the formation of cytochromes aa_3 and/or b.

Mitochondriogenesis is studied normally in organisms that exhibit easily initiated, synchronous biosynthetic events. Yeast cultures are commonly used because of their ability to initiate the production of mitochondria within 30 min after the conversion from anaerobic to aerobic conditions or after the removal of glucose from the culture medium. However, yeasts lack endocrine regulation of their biosynthetic processes. Therefore, although yeasts demonstrate basic biosynthetic events of mitochondriogenesis, they may have regulations over these events different from those found in the higher animals, in which hormones may play a significant role.

Mitochondriogenetic studies are difficult to do in higher animals with defined hormones because tissue growth is rarely rapid or synchronous. Vertebrate mitochondriogenesis has been studied during regeneration of the liver following hepatectomy, but normal hormonal influences and responses are uncertain under these traumatic circumstances. Therefore, the fat body of *Blaberus* offers a unique system for studying natural hormonal influences on mitochondriogenesis in higher animals.

Two central questions remain. First, what is the physiological significance of endocrine regulation of cytochromes in the fat body mitochondria? Second, how does CGF stimulate cytochrome synthesis? At present, we are unable to answer either of these questions, but both are under investigation.

We speculate that the physiological purpose of fat body mitochondriogenesis is to increase the energy-generating capacity of the tissue to meet adult biosynthetic demands. This probably consists of producing the hemolymph metabolites necessary for muscular activity, since $CC^- + CA^-$ animals take longer to recover from exhaustion (Shepard and Keeley, 1973); in females, the fat body energy needs are considerable for the synthesis of the yolk precursors during vitellogenesis. We have recently confirmed that in the fat body of female *Blaberus* the mitochondrial respiration also shows a CC-dependent, threefold increase during early adult life (McKercher and Keeley, unpublished results).

The action of CGF at the subcellular level is at present unknown. CGF may stimulate fat body metabolism, thus increasing the energy demands of the tissue and activating mitochondriogenesis by an unknown feedback mechanism. However, the specificity of CGF on the synthesis of cytochromes $aa_3 + b$ argues against a secondary effect on general mitochondriogenesis. If the CGF effect on mitochondria was of a secondary nature, then all mitochondriogenetic processes should be activated and the synthesis of cytochromes $c + c_1$ should also respond. Therefore, we currently feel that CGF has a specific regulatory action on cytochrome synthesis.

CGF may regulate the formation of cytochromes $aa_3 + b$ at the level of heme synthesis. CGF may either stimulate the enzymes in the general heme synthesis pathway (ALA dehydratase, ferrochelatase, etc.) or it may regulate the balance between the synthesis of specific hemes. For example, either protoheme (heme b) is converted to heme a (Sinclair *et al.*, 1967), or protoporphyrin IX may be a common precursor for both protoheme and heme a (Keyhani and Keyhani; 1976). In either case, CGF may affect the enzymes that determine the formation of a specific heme, i.e., heme a or b or both.

Alternatively, CGF may regulate the synthesis of the cytochrome apoproteins. Apoprotein production also facilitates increased incorporation of the hemes formed from [^{14}C]-ALA. Studies show that the production of apoproteins is complex, since both cytochromes aa_3 and b reportedly have seven subunits (Ebner *et al.*, 1973; Marjanen and Ryrie,

1976). Four of the subunits are translated on cytoplasmic ribosomes, and three are translated on mitoribosomes. Assuming the apoproteins of *Blaberus* and yeasts are similar, CGF may stimulate either translation system to produce its subunits.

The isolation and characterization of CGF, which are now in progress, combined with the fat body model system for mitochondriogenesis in *Blaberus* will provide basic information of use to those interested in both general cellular and insect biochemistry. The system may elucidate regulations on mitochondriogenesis in the cells of higher animals, and it may demonstrate a new endocrine control of fat body metabolism in insects based on mitochondrial integrity and function.

ACKNOWLEDGMENTS. The author wishes to thank Dr. F. W. Plapp for critically reviewing this manuscript; Dr. E. L. Thurston, director of the Electron Microscopy Center of Texas A&M University for his invaluable assistance in the studies on fat body structure; and Drs. Herbert Roller and Karl Dahm for their generous contributions of the juvenile hormones and ecdysone tested in these studies. This research was supported by the Texas Agricultural Experiment Station and grants from NSF (PCM 74-03606) and NIH (5 ROI AI 15190-02 TMP).

References

Basford, R.E., Tisdale, H.D., Glenn, L., and Green, D.E., 1957, Studies on the terminal electron transport system. VII. Further studies on the succinic dehydrogenase complex, *Biochim. Biophys. Acta* **24**:107.

Beattie, D.S., 1971, The possible relationship between heme synthesis and mitochondrial biogenesis, *Arch. Biochem. Biophys.* **147**:136.

Beattie, D.S., and Stuchell, R.N., 1970, Studies on the induction of hepatic δ-aminolevulinic acid synthetase in rat liver mitochondria, *Arch. Biochem. Biophys.* **139**:291.

Chan, S.K., Reibling, A., Mahaffey, W.L., and Lin, C.C., 1973, The δ-aminolevulinic acid synthetase activity and the effect of exogenous δ-aminolevulinate on the synthesis of cytochrome *c* in the thoracic muscles of the tobacco hornworm during adult development, *Biochim. Biophys. Acta* **329**:251.

Chaudhary, K.D., and Malhotra, L., 1974, Effet de l'ecdysone sur l'incorporation *in vivo* des acides amines dans les proteines des fractions mitochondriale et postmitochondriale de muscle thoracique de criquet *Schistocerca gregaria*, *Rev. Can. Biol.* **33**:161.

Chefurka, W., 1978, Sesquiterpene juvenile hormones: Novel uncouplers of oxidative phosphorylation, *Biochem. Biophys. Res. Commun.* **83**:571.

Clarke, K.U., and Baldwin, R.W., 1960, The effect of insect hormones and of 2:4-dinitrophenol on the mitochondrion of *Locusta migratoria* L., *J. Insect Physiol.* **5**:37.

Clements, A.N., 1963, *The Physiology of Mosquitoes*, Pergamon Press, Oxford.

Ebner, E., Mason, T.L., and Schatz, G., 1973, Mitochondrial assembly in respiratory-

deficient mutants of *Saccharomyces cerevisiae*. II. Effect of nuclear and extrachromosomal mutations on the formation of cytochrome *c* oxidase, *J. Biol. Chem.* **248:**5369.
Firstenberg, D.E., and Silhacek, D.L., 1973, Juvenile hormone regulation of oxidative metabolism in isolated insect mitochondria, *Experientia* **29:**1420.
Frontali, N., and Gainer, H., 1977, Peptides in invertebrate nervous systems, in: *Peptides in Neurobiology* (H. Gainer, ed.), pp. 259–294, Plenum Press, New York.
Granick, S., and Kappas, A., 1967, Steroid induction of porphyrin synthesis in liver cell culture. 1. Structural basis and possible physiological role in the control of heme formation, *J. Biol. Chem.* **242:**4587.
Granick, S., and Urata, G., 1963, Increase in activity of δ-aminolevulinic acid synthetase in liver mitochondria induced by feeding of 3,5-dicarbethoxy-1,4-dihydrocollidine, *J. Biol. Chem.* **238:**821.
Gruetzmacher, M.C., and Keeley, L.L., 1981, Primary biochemical events during diapause termination in *Heliothis zea*: The role of mitochondrial development, *Insect Biochem.*, in review.
Hayes, T.K., and Keeley, L.L., 1981, Cytochromogenic factor: A newly-discovered neuroendocrine agent stimulating mitochondrial cytochrome synthesis in the insect fat body, *Gen. Comp. Endocrinol.*, in press.
Keeley, L.L., 1970, Insect fat body mitochondria: Endocrine and age effects on respiratory and electron transport activities, *Life Sci.* **9:**1003.
Keeley, L.L., 1971, Endocrine effects on the biochemical properties of fat body mitochondria from the cockroach *Blaberus discoidalis*, *J. Insect Physiol.* **17:**1501.
Keeley, L.L., 1972, Biogenesis of mitochondria: Neuroendocrine effects on the development of respiratory functions in fat body mitochondria of the cockroach, *Blaberus discoidalis*, *Arch. Biochem. Biophys.* **153:**8.
Keeley, L.L., 1973, Characterization of insect fat body mitochondria isolated by a rapid procedure, *Comp. Biochem. Physiol.* (B) **46:**147.
Keeley, L.L., 1978a, Development and endocrine regulation of mitochondrial cytochrome biosynthesis in the insect fat body. I. δ-Aminolevulinic acid synthase, *Arch. Biochem. Biophys.* **187:**78.
Keeley, L.L., 1978b, Development and endocrine regulation of mitochondrial cytochrome biosynthesis in the insect fat body. II. δ-[^{14}C]Aminolevulinic acid incorporation, *Arch. Biochem. Biophys.* **187:**87.
Keeley, L.L., and Friedman, S., 1967, Corpus cardiacum as a metabolic regulator in *Blaberus discoidalis* Serville (Blattidae). Long-term effects of cardiacectomy on whole body and tissue respiration and trophic metabolism, *Gen. Comp. Endocrinol.* **8:**129.
Keeley, L.L., and Friedman, S., 1969, Effects of long-term cardiacectomy–allatectomy on mitochondrial respiration in the cockroach, *Blaberus discoidalis*, *J. Insect Physiol.* **15:**509.
Keeley, L.L., and Waddill, V.H., 1971, Insect hormones: Evidence for a neuroendocrine factor affecting respiratory metabolism. *Life Sci.* **10**(II)**:**737.
Keyhani, J., and Keyhani, E., 1976, Porphyrin *a* as a precursor of heme *a* in *Candida utilis*, *FEBS Lett.* **70:**118.
Larsen, W.J., 1970, Genesis of mitochondria in insect fat body, *J. Cell Biol.* **47:**373.
Larsen, W.J., 1976, Cell remodeling in the fat body of an insect, *Tissue & Cell* **8:**73.
Locke, M., and Collins, J.V., 1965, The structure and formation of protein granules in the fat body of an insect, *J. Cell Biol.* **26:**857.
Luscher, M., 1968, Hormonal control of respiration and protein synthesis in the fat body of the cockroach *Nauphoeta cinerea* during oocyte growth, *J. Insect Physiol.* **14:**499.
Luscher, M., and Leuthold, R., 1965, Uber die hormonale Beeinflussung des respirato-

rischen Stoffwechsels bei der Schabe *Leucophaea maderae* (F.), *Rev. Suisse Zool.* **72**:618.

Mannix, J.J., and Keeley, L.L., 1980, Age and endocrine effects on fat body metabolite composition in adult male *Blaberus discoidalis* cockroaches, *J. Exp. Zool.* **212**:113.

Marjanen, L.A., and Ryrie, I.J., 1976, The biogenesis of a cytochrome *b* complex in yeast mitochondria. Sites of translation of the protein components, *Arch. Biochem. Biophys.* **172**:679.

Marx, R., 1971, Morphological changes in the fat body mitochondria during larval development of the blowfly, *Calliphora erythrocephala* Meigen, *Cytobiologie Z. Exp. Zellforsch.* **3**:417.

Minks, A.K., 1967, Biochemical aspects of juvenile hormone action in the adult *Locusta migratoria, Arch. Neerl. Zool.* **17**:175.

Muller, H.P., and Engelmann, F., 1968, Studies on the endocrine control of metabolism in *Leucophaea maderae* (Blattaria). II. The effect of the corpora cardiaca on fat-body respiration, *Gen. Comp. Endocrinol.* **11**:43.

Osanai, M., and Rembold, H., 1970, Biosynthesis of cytochrome *c*. I. Incorporation *in vivo* of radioactive iron ($^{59}Fe^{2}$) and of [^{14}C]lysine into cytochrome *c* of the honeybee, *Hoppe-Seyler's Z. Physiol. Chem.* **351**:643.

Osanai, M., and Rembold, H., 1974, Biosynthesis of cytochrome *c* IV. The activities of 5-aminolevulinate and porphobilinogen synthase at different stages of development of the honeybee, *Hoppe-Seyler's Z. Physiol. Chem.* **355**:327.

Osanai, M., and Rembold, H., 1975, Biosynthesis of cytochrome *c*. V. Concentrations of 5-aminolevulinate, aminoacetone and porphobilinogen at different stages of development of the honeybee, *Hoppe-Seyler's Z. Physiol. Chem.* **356**:15.

Pfeiffer, I.W., 1945, Effect of the corpora allata on the metabolism of adult female grasshoppers, *J. Exp. Zool.* **99**:183.

Pratt, G.E., Tobe, S.S., Weaver, R.J., and Finney, J.R., 1975, Spontaneous synthesis and release of C16 juvenile hormone by isolated corpora allata of female locust *Schistocerca gregaria* and female cockroach *Periplaneta americana, Gen. Comp. Endocrinol.* **26**:478.

Sagesser, H., 1960, Uber die Wirkung der Corpora Allata auf den Sauerstoffverbrauch bei der Schabe *Leucophaea maderae* (F.), *J. Insect Physiol.* **5**:264.

Saltzgaber-Muller, J., and Schatz, G., 1978, Heme is necessary for the accumulation and assembly of cytochrome *c* oxidase subunits in *Saccharomyces cerevisiae, J. Biol. Chem.* **253**:305.

Scholnick, P.L., Hammaker, L.E., and Marver, H.S., 1972, Soluble δ-aminolevulinic acid synthetase of rat liver. II. Studies related to the mechanism of enzyme action and hemin inhibition, *J. Biol. Chem.* **247**:4132.

Shappirio, D.G., 1960, Oxidative enzymes and the injury metabolism of diapausing cecropia silkworms, *Ann. N.Y. Acad. Sci.* **89**:537.

Shepard, M., and Keeley, L.L., 1973, Circadian rhythmicity and capacity for enforced activity in the cockroach, *Blaberus discoidalis*, after cardiacectomy–allatectomy, *J. Insect Physiol.* **18**:595.

Sinclair, P., White, D.C., and Barrett, J., 1967, The conversion of protoheme to heme *a* in *Staphylococcus, Biochim. Biophys. Acta* **143**:427.

Slama, K., 1964, Hormonal control of respiratory metabolism during growth, reproduction, and diapause, in female adults of *Pyrrhocoris apterus* L. (Hemiptera), *J. Insect Physiol.* **10**:283.

Slama, K., 1965, Effect of hormones on the respiration of body fragments of adult *Pyrrhocoris apterus* L. (Hemiptera), *Nature (London)* **205**:416.

Slama, K., and Kryspin-Sorensen, I., 1979, Hypermetabolic response induced by juvenile hormone analogues in an insect, *Z. Naturforsch.* **34c**:599.

Soslau, G., Stotz, E.H., and Lockshin, R.A., 1971, Stimulation of cytochrome *c* synthesis in the developing polyphemus moth by δ-aminolevulinic acid, *Biochemistry* **10**:3296.

Thomsen, E., 1949, Influence of the corpus allatum on the oxygen consumption of adult *Calliphora erythrocephala* Meig., *J. Exp. Biol.* **26**:137.

Thomsen, E., and Hamburger, K., 1955, Oxygen consumption of castrated females of the blowfly, *Calliphora erythrocephala* Meig., *J. Exp. Biol.* **32**:692.

Thomsen, E., and Thomsen, M., 1974, Fine structure of the fat body of the female *Calliphora erythrocephala* during the first egg maturation cycle, *Cell Tissue Res.* **152**:193.

Tsuyama, S., and Miura, K., 1979, Structural and enzymic changes in fat body mitochondria of the blowfly, *Aldrichina grahami*, during larval growth, *Insect Biochem.* **9**:435.

Wiens, A.W., and Gilbert, L.I., 1965, Regulation of cockroach fat body metabolism by the corpus cardiacum *in vitro*, *Science* **150**:614.

Wigglesworth, V.B., 1935, Function of the corpus allatum in insects, *Nature* (*London*) **136**:338.

Williams, J.N., 1964, A method for the simultaneous quantitative estimation of cytochromes a, b, c_1 and c in mitochondria. *Arch. Biochem. Biophys.* **107**:537.

Woods, R.A., Sanders, H.K., Briquet, M., Foury, F., Drysdale, B.-E., and Mattoon, J.R., 1975, Regulation of mitochondrial biogenesis: Enzymatic changes in cytochrome-deficient yeast mutants requiring δ-aminolevulinic acid, *J. Biol. Chem.* **250**:9090.

Index

GPSR Compliance

The European Union's (EU) General Product Safety Regulation (GPSR) is a set of rules that requires consumer products to be safe and our obligations to ensure this.

If you have any concerns about our products, you can contact us on ProductSafety@springernature.com

In case Publisher is established outside the EU, the EU authorized representative is:

Springer Nature Customer Service Center GmbH
Europaplatz 3
69115 Heidelberg, Germany

Batch number: 10844675

Printed by Printforce, the Netherlands